2/01

Digital Neuroanatomy

Digital Neuroanatomy

George R. Leichnetz
Department of Anatomy and Neurobiology
School of Medicine,
Virginia Commonwealth University

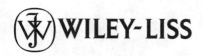**WILEY-LISS**

A John Wiley & Sons, Inc., Publication

Published by John Wiley & Sons, Inc., Hoboken, New Jersey
Published simultaneously in Canada

For general information on our other products and services or for technical support, please contact our Customer Care Department within the United States at (800) 762-2974, outside the United States at (317) 572-3993 or fax (317) 572-4002.

Wiley also publishes its books in a variety of electronic formats. Some content that appears in print may not be available in electronic formats. For more information about Wiley products, visit our web site at www.wiley.com.

Library of Congress Cataloging-in-Publication Data is available.

ISBN-10 0-470-04000-9
ISBN-13 978-0-470-04000-3

Printed in the United States of America

10 9 8 7 6 5 4 3 2 1

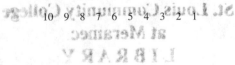

To my wife and best friend, Athalie, whose faith, love,
and steadfast support has inspired, encouraged,
and sustained me and our family.

Contents

Preface

This computer-based program is intended to present what is considered that every medical student should know about neuroanatomy taught in wet laboratory sessions in a first-year medical neuroscience course. The program does not purport to be an exhaustive presentation of this subject material as this is no longer feasible with the constraints of time in the current trend to compress medical school courses. Our medical students have a second-year course in which the pathology of the brain is discussed more extensively.

The study begins with a presentation of essential light-microscopic and electron microscopic neurohistology that the student needs to know in order to understand the substrate of cellular and molecular neuroscience underlying the bioelectrical and neurochemical functioning of the central nervous system (CNS). Although medical students are exposed to the skull and meninges in gross anatomy, they are reviewed in this program so the student is reminded of the bony environment of the brain, the foramina traversed by cranial nerves, and compartmentalization of the cranial vault resulting from meningeal partitions, to gain an understanding of the physical impediments within the cranium for the growth and expansion of space-occupying tumors. Next, the program provides a study of the gross and sectional anatomy of the CNS to help the student develop an adequate structural vocabulary for the subsequent study of the connections and functions of the brain and the consequences of injury or disease. The introduction to brain imaging is intended primarily to show the correlation of magnetic resonance imaging (MRI) with sectional brain anatomy. Images of lesions and tumors in MRIs are provided to motivate the student to determine what brain structures are affected in the pathology, and lead them to a prediction and diagnosis of associated clinical neurological deficits.

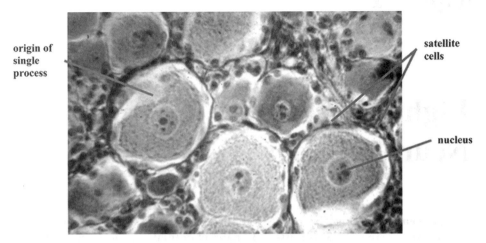

origin of
single
process

satellite
cells

nucleus

Figure 1.1b   Unipolar neurons in dorsal root ganglion.

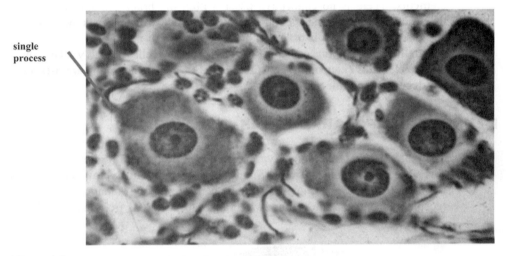

single
process

Figure 1.1c   Unipolar neurons in dorsal root ganglion, silver.

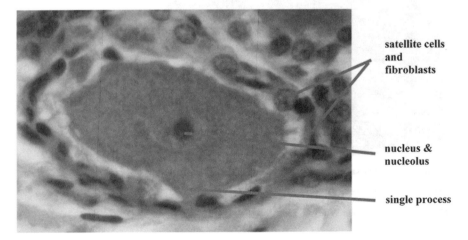

satellite cells
and
fibroblasts

nucleus &
nucleolus

single process

Figure 1.1d   Unipolar neuron in dorsal root ganglion Hematoxylin and eosin (H & E). Oil immersion.

large blood vessels. They have a round cell body with central nucleus and have a single major process that comes off the cell body. Since the single process immediately divides into a peripheral process that goes out to the receptor and a central process that carries the sensory information into the CNS, unipolar neurons are sometimes referred to as "pseudounipolar." While both processes are histologically "axons," the peripheral process functions like a dendrite, conducting sensory information toward the cell body. Their cell bodies are found in **dorsal root ganglia** and **cranial sensory ganglia**. The larger cell bodies are somatic, while the smaller ones are associated with visceral sensation. Unipolar neurons are derived embryologically from neural crest.

Bipolar neurons are special sensory (associated with special senses). Their cell bodies are found in the retina, vestibular, and cochlear ganglia and the olfactory epithelium. They typically have an ovoid cell body and two processes: a peripheral and central process. While the retina develops as an outgrowth of the embryonic diencephalon, the bipolar neurons of the vestibular and cochlear ganglia and the olfactory epithelium develop from specialized regions of neuroepithelium on the surface of the embryo known as placodes.

Bipolar neurons are present in three layers of the **retina** (Fig. 1.2a,b). Neurons in the outer nuclear layer of the retina have **rods** and **cones** on their peripheral processes, which contain visual pigments and are receptive to light/dark or color, respectively. Ganglion cells give rise to the axons of the optic nerve.

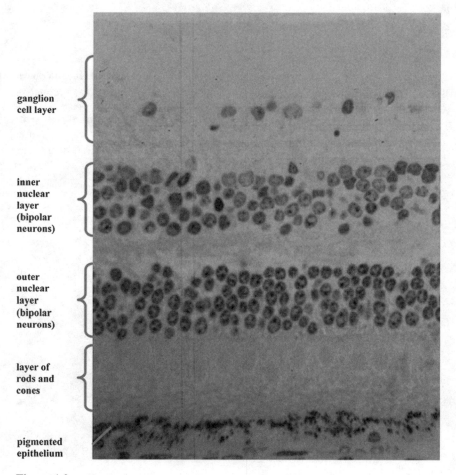

ganglion
cell layer

inner
nuclear
layer
(bipolar
neurons)

outer
nuclear
layer
(bipolar
neurons)

layer of
rods and
cones

pigmented
epithelium

Figure 1.2a   **Bipolar neurons in the retina.**

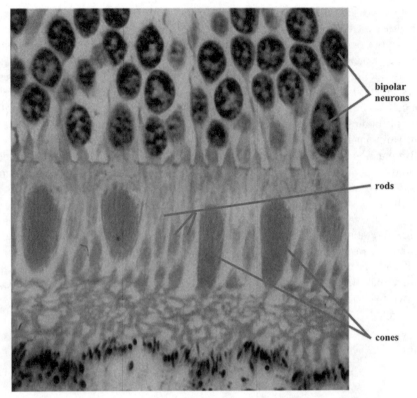

Figure 1.2*b*   **Rods and cones on the peripheral processes of bipolar neurons in the outer nuclear layer of the retina.**

Bipolar neurons are also found in the **cochlear** (spiral) **ganglion in the cochlea of the inner ear** (Fig. 1.2c,d). Their peripheral processes end in auditory receptor hair cells in the organ of Corti, and their central processes join the auditory division of the vestibulocochlear nerve [cranial nerve (CN. VIII)] to reach the brainstem. Bipolar neurons are also

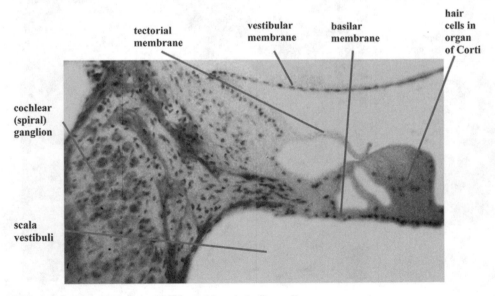

Figure 1.2*c*   **Bipolar neurons in the cochlear (spiral) ganglion.**

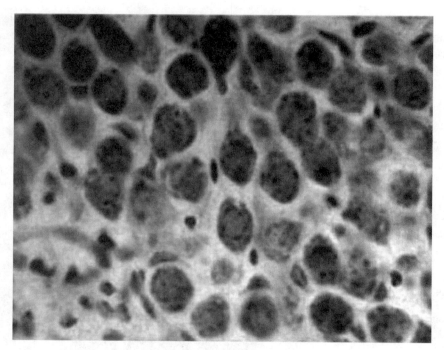

Figure 1.2d   Bipolar neurons in cochlear ganglion.

found in the *vestibular ganglion* (Fig. 1.2e) associated with receptors in the ampullae of the semicircular canals and maculae of the otolith organs (saccule and utricle). Their central processes join the vestibular division of the vestibulocochlear nerve (CN. VIII).

Multipolar neurons in the PNS are found in autonomic ganglia (e.g., sympathetic chain ganglia, preaortic ganglia) and the adrenal medulla.

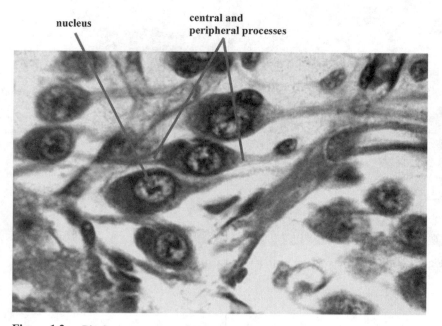

Figure 1.2e   Bipolar neurons in vestibular ganglion.

CNS multipolar neurons are of a wide variety of size and shape, for example, motor neurons of the ventral horn of the spinal cord (Fig 1.3a–c) pyramidal cells of the cerebral cortex (Fig. 1.3d,e), and Purkinje cells of the cerebellar cortex (Fig. 1.3f–h).

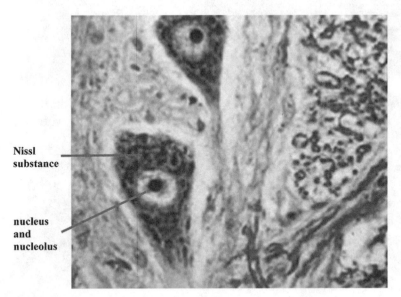

Figure 1.3*a*   Motor neurons in spinal cord ventral horn. Luxol fast blue/cresyl violet.

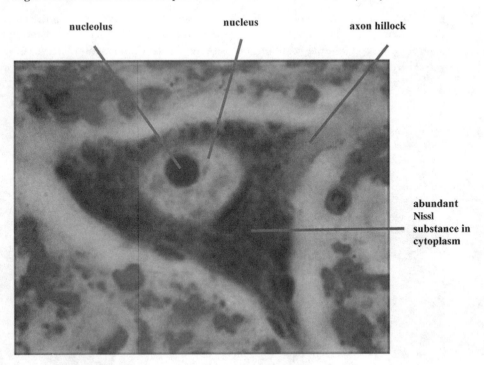

Figure 1.3*b*   Multipolar motor neuron in spinal cord ventral horn.

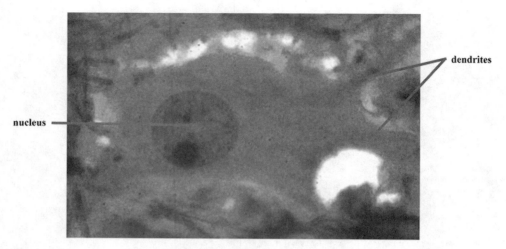

nucleus

dendrites

Figure 1.3c   Anterior horn motor neuron. Silver.

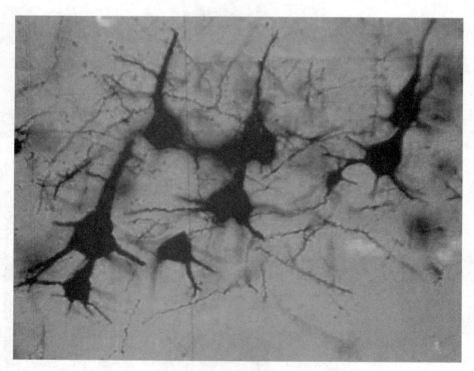

Figure 1.3d   Multipolar pyramidal neurons in cerebral cortex. Silver.

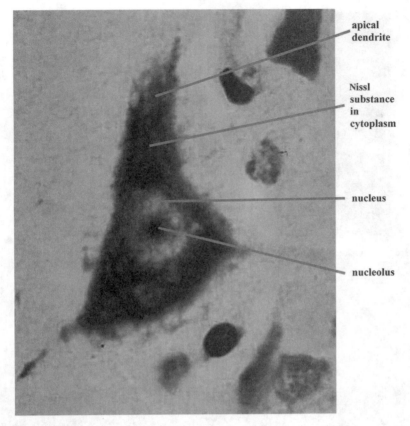

apical
dendrite

Nissl
substance
in
cytoplasm

nucleus

nucleolus

Figure 1.3e   Multipolar pyramidal neuron. Luxol fast blue/cresyl violet.

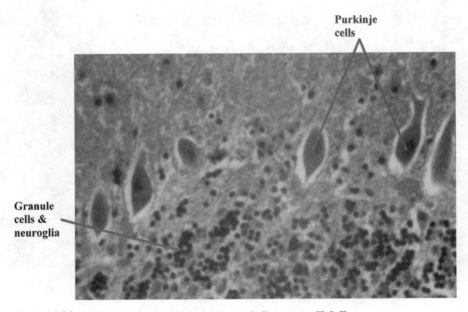

Purkinje
cells

Granule
cells &
neuroglia

Figure 1.3f   Multipolar Purkinje neurons in cerebellar cortex. H & E.

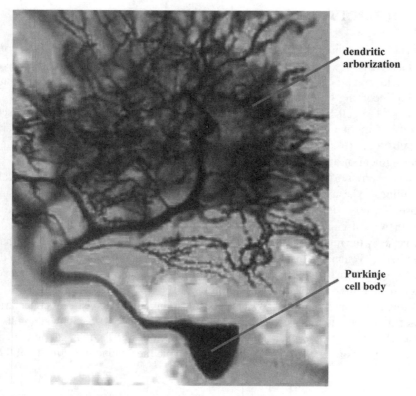

dendritic
arborization

Purkinje
cell body

Figure 1.3g   Multipolar Purkinje neuron in cerebellar cortex. Silver.

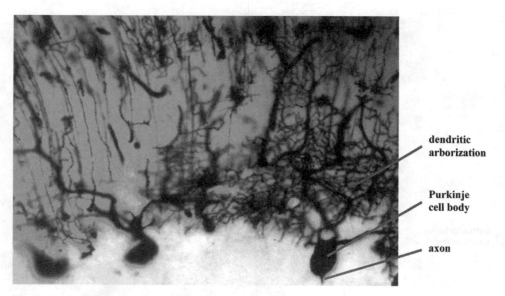

dendritic
arborization

Purkinje
cell body

axon

Figure 1.3h   Multipolar cerebellar Purkinje neurons. Silver.

TYPICAL MULTIPOLAR NEURON

A typical neuron, such as a multipolar neuron of the ventral horn of the spinal cord, consists of a cell body (soma or perikaryon) with multiple tapering processes called **dendrites**, and a single long process of uniform diameter, the **axon.** The dendrites act like "antennae" receiving incoming bioelectrical signals, whereas the axon conducts the neural impulse away from the cell body and carries the signal to another neuron or to a muscle. The cell body of the neuron contains the **nucleus, nucleolus, mitochondria**, and abundant **Nissl substance** (rough endoplasmic reticulum, RER), **Golgi complex, lysosomes**, lipofuscin (pigment of age), and neuromelanin.

The region of origin of the axon from the soma lacks Nissl substance and is called the **axon hillock**. The axon terminates in relation to another neuron (or on a muscle) in a **synapse.** The axon terminal or bouton is not directly apposed to the postsynaptic membrane; there is a gap (**synaptic cleft**) between them. The impulse is conducted across the synapse by the release of neurochemical transmitters from **synaptic vesicles** that selectively affect ion channels in the postsynaptic membrane (**ionotropic**), resulting in depolarization (excitation) or hyperpolarization (inhibition), or have a neuromodulatory effect through second messengers (**metabotropic**).

Axons are typically insulated by a fatty sheath of myelin that consists of concentric wrappings of membranous extensions from oligodendrocytes (CNS) or Schwann cells (PNS). Myelination occurs in internodal segments between **nodes of Ranvier** along the course of an axon. The thicker the myelin sheath, the more rapid the conduction velocity of the nerve.

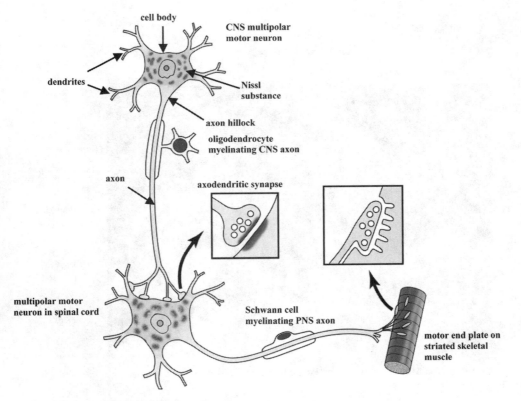

Figure 1.4   Typical multipolar neurons.

GRAY MATTER VS. WHITE MATTER

Gray matter in the CNS contains the cell bodies of neurons that are not surrounded with myelin. A cluster of cell bodies in the CNS is referred to as a **nucleus** (e.g., caudate nucleus, hypoglossal nucleus (Fig. 1.5a), and layers of cell bodies on the surface of the

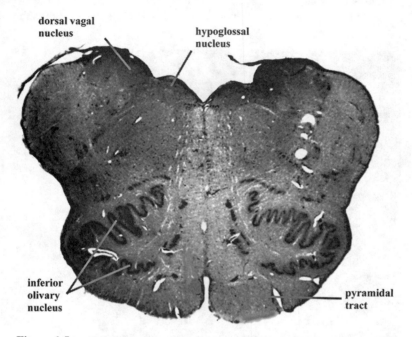

dorsal vagal nucleus

hypoglossal nucleus

inferior olivary nucleus

pyramidal tract

Figure 1.5a   Cytoarchitecture. Medulla stained with cresyl violet shows nuclei (gray matter).

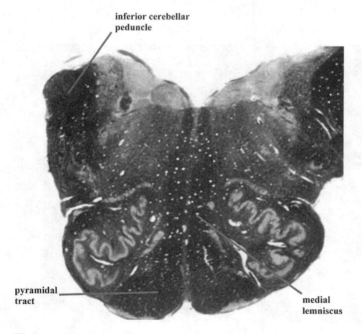

inferior cerebellar peduncle

pyramidal tract

medial lemniscus

Figure 1.5b   Myeloarchitecture. Medulla stained for myelin shows tracts (white matter).

cerebrum and cerebellum are referred to as a **cortex** (e.g., cerebral cortex, cerebellar cortex). Basophilic dyes, like cresyl violet, stain the Nissl substance in the cell bodies of neurons in neurons (or cortex) most darkly.

  White matter in the CNS contains predominantly **myelinated axons** and is stained most darkly with myelin stains, like Weigert-Pal, and contains groups of axons that run together in **tracts** or **fasciculi** (Fig. 1.5b). The large regions of white matter of the outer portion of the spinal cord contain numerous tracts (Fig. 1.6a,b) and are referred to as

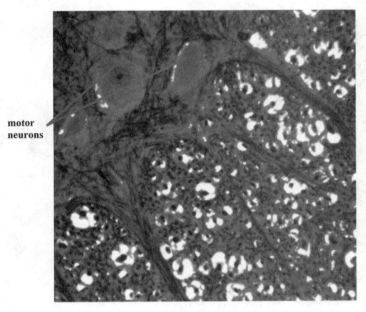

motor
neurons

Figure 1.6*a*   Spinal cord ventral horn and adjacent white matter.

myelinated axons

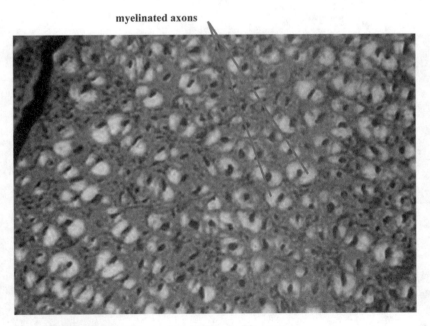

Figure 1.6*b*   Myelinated axons in spinal cord white matter.

columns or **funiculi** (singular, funiculus). In the cerebrum, fascicles of myelinated fibers may surround or course through a structure forming **capsules** (e.g., internal capsule, external capsule) or **laminae** (e.g., internal medullary lamina of the thalamus). These structures are made up of myelinated axons, not connective tissue.

NEUROGLIA: PROTOPLASMIC AND FIBROUS ASTROCYTES, OLIGODENDROCYTES, AND MICROGLIA

CNS neuroglia are primarily of two types: **astrocytes** and **oligodendrocytes**, both of which are derived from embryonic neural ectoderm of the neural tube. A third type, **microglia**, is derived from embryonic mesoderm and transforms into phagocytes with brain injury. **Astrocytes** are positioned between neurons and capillaries and provide metabolic support for neurons. They surround synapses and can take up excess excitatory neurotransmitters to prevent neuronal damage. In the embryonic CNS, astrocytes also form glial planes that guide the structural development of the brain. Astrocytes send processes to capillaries, **perivascular end feet**, which cover at least 80% of capillary surface and augment the blood–brain barrier. The **blood–brain barrier** is predominantly a property of the capillary endothelium that selectively prevents substances from entering the brain. It has tight junctions (zonula occludens) that make it selectively permeable, preventing passive diffusion of large molecules.

Protoplasmic astrocytes are found predominantly in gray matter (Fig. 1.7a–c). In an H & E or cresyl violet-stained section they have nuclei that are larger and euchromatic, whereas oligodendrocytes have smaller spheroidal, densely stained nuclei, and are heterochromatic. In a silver-stained section protoplasmic astrocytes have a mossy ("tumbleweed") appearance.

Fibrous astrocytes are found predominantly in white matter (Fig. 1.8a–c) and have fewer, longer, less highly branched processes.

protoplasmic
astrocytes

pyramidal
neurons

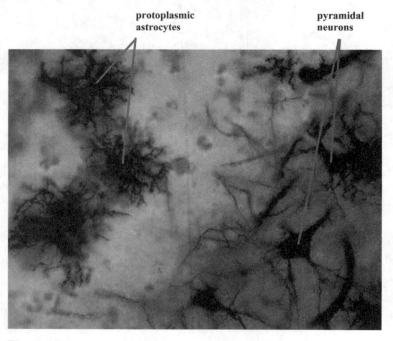

Figure 1.7a   Protoplasmic astrocytes in cerebral cortex. Silver.

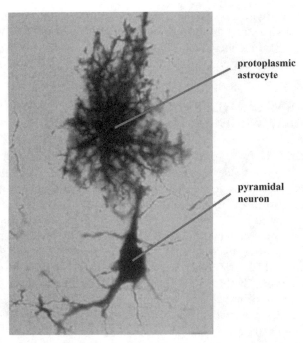

Figure 1.7*b*   Protoplasmic astrocyte with pyramidal neuron. Silver.

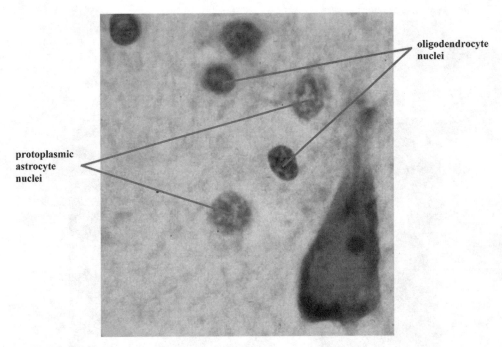

Figure 1.7*c*   Cerebral cortex. Pyramidal cell with protoplasmic astrocyte and oligodendrocyte nuclei. Luxol fast blue/cresyl violet.

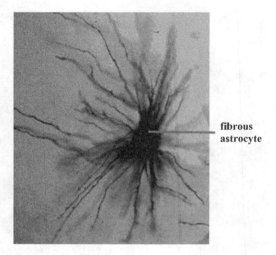

Figure 1.8*a*   Fibrous astrocytes in subcortical white matter. Silver.

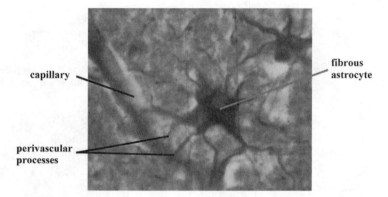

Figure 1.8*b*   Fibrous astrocyte with perivascular end feet to capillary.

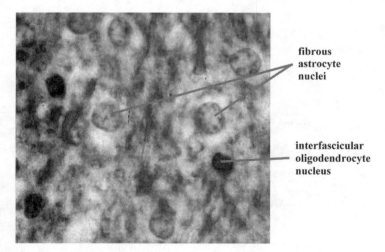

Figure 1.8*c*   Subcortical white matter. Cresyl violet (CV). Fibrous astrocyte and oligodendrocyte nuclei.

oligodendrocytes

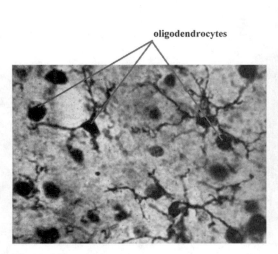

Figure 1.9*a* Oligodendrocytes.

interfascicular oligodendrocyte nuclei

fibrous astrocyte nucleus

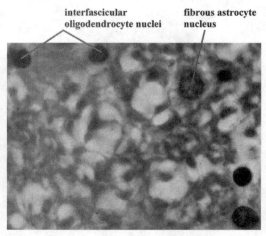

Figure 1.9*b* Oligodendrocyte and astrocyte nuclei in spinal cord white matter. H & E. Oil immersion.

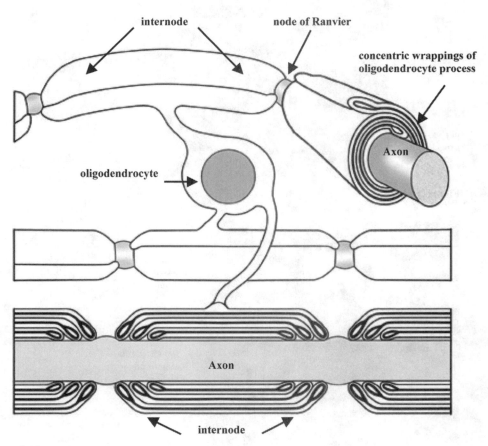

Figure 1.10*a* Oligodendrocytes myelinate CNS axons.

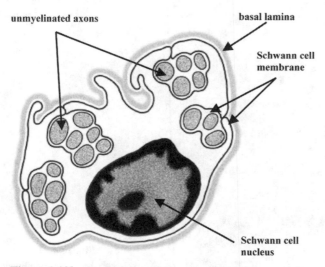

unmyelinated axons

basal lamina

Schwann cell membrane

Schwann cell nucleus

Figure 1.10*b*   **Unmyelinated PNS nerves where Schwann cell processes surround several axons without producing concentric lamellae.**

Oligodendrocytes (Fig. 1.9a,b) myelinate CNS axons, whereas Schwann cells myelinate PNS axons. While oligodendrocytes may send out multiple processes that myelinate as many as 50 internodal segments, Schwann cells myelinate a single internode of a peripheral axon (Fig. 1.10a). In some cases a Schwann cell process may surround a group of axons without individually wrapping them (hence, no individual myelin sheath), which are therefore called **unmyelinated nerves** (Fig. 1.10b).

In an H & E stain at the LM level, **oligodendrocytes** have small spheroidal, densely stained heterochromatic nuclei (Fig. 1.11a,b).

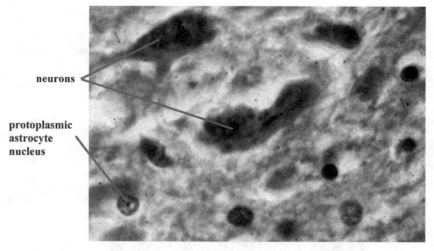

neurons

protoplasmic astrocyte nucleus

Figure 1.11*a*   **Protoplasmic astrocyte and oligodendrocyte nuclei in spinal cord gray matter. H & E.**

pyramidal
neurons

perineuronal
oligodendrocyte
nucleus

protoplasmic
astrocyte nuclei

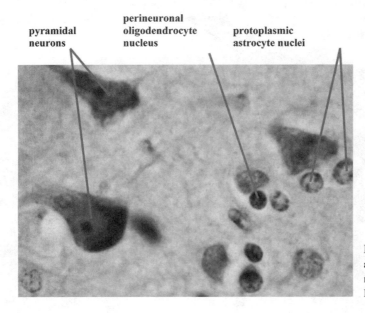

Figure 1.11*b* Protoplasmic astrocyte and oligodendrocyte nuclei in cerebral cortex. Luxol fast blue/cresyl violet.

PERIPHERAL NERVES: EPINEURIUM, PERINEURIUM, AND ENDONEURIUM

In the PNS, peripheral nerves are covered with connective tissue. **Epineurium** is loose connective tissue and surrounds the entire nerve (Fig. 12.a,b). The **perineurium** is denser connective tissue and surrounds individual fascicles of nerve fibers within the nerve. The **endoneurium** surrounds individual nerve fibers outside the myelin sheath. In an injured nerve where the axon and its myelin sheath degenerate (Wallerian degeneration), the remaining endoneurial tubes leave channels through which the peripheral nerve

perineurium

nerve
fascicle

epineurium

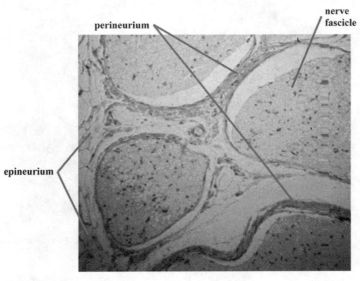

Figure 1.12*a* Peripheral nerve in cross section. H & E.

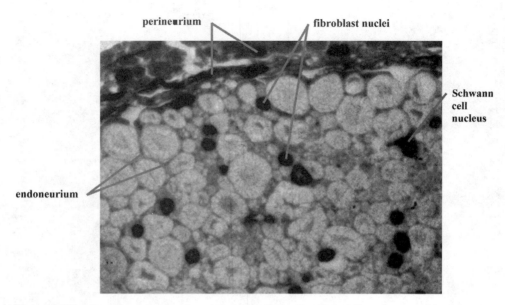

Figure 1.12*b*   Fascicle of peripheral nerve in cross section showing myelinated axons with endoneurium, fibroblast and Schwann cell nuclei. H & E. Oil immersion.

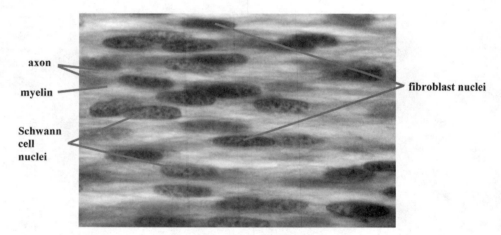

Figure 1.12*c*   Peripheral nerve longitudinal section showing fibroblast and Schwann cell nuclei. H & E.

can regenerate. Within the nerve **fibroblast nuclei** that synthesize the connective tissue collagen fibers are elongated, whereas the **Schwann cell nuclei** that produce the myelin are plump (Fig. 1.12c).

Chapter 2

Electron-Microscopic (EM) Neurohistology

NEURONAL SOMA (CELL BODY) AND ORGANELLES (RER, GOLGI COMPLEX, MITOCHONDRIA, LYSOSOMES, LIPOFUSCIN)

The **neuron cell body** (soma) contains the **nucleus** and **nucleolus**, and its cytoplasm contains abundant organelles, including **rough endoplasmic reticulum** (which represents

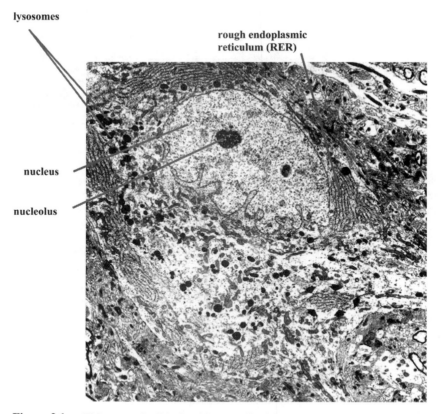

Figure 2.1a EM neuronal cell body with organelle-rich cytoplasm, prominent rough endoplasmic reticulum (RER, Nissl substance), lysosomes, and mitochondria.

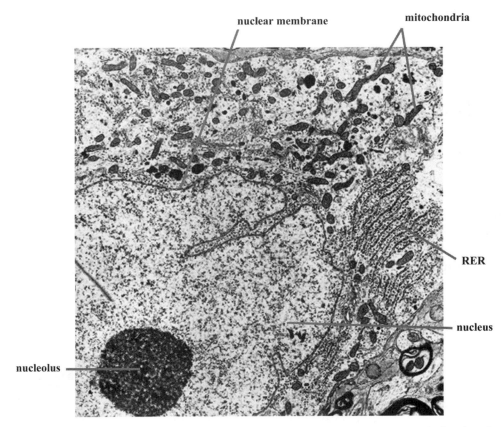

Figure 2.1*b*   **EM neuronal soma with nucleus, nucleolus, and cytoplasm with rough endoplasmic reticulum (RER), mitochondria, and lysosomes.**

Nissl substance at the LM level) **mitochondria**, **lysosomes**, and **lipofuscin** (pigment of age) (Fig. 2.1a–d). The **Golgi complex** is in a perinuclear location and is contiguous with the RER (Fig 2.1d). It concentrates neuronal products into vesicles.

DENDRITES: NEUROTUBULES AND NEUROFILAMENTS

Dendrites are tapering processes that come off the cell body. The proximal portion of primary dendrites may be seen in continuity with the soma and may contain some RER and mitochondria in addition to the neuron's principal cytoskeletal elements, the **neurotubules** (microtubules) and **neurofilaments** (Fig. 2.2a,b). Smaller secondary or tertiary dendritic profiles often represent sites for the termination of axons from other neurons, **axodendritic synapses**.

DENDRITIC SPINES

Secondary and tertiary dendrites are usually studded with **dendritic spines** (Fig. 2.3a,b), which are chalice-like structures that increase dendritic surface area and offer sites for additional input to the neuron through **axodendritic synapses** (Fig. 2.4a).

lipofuscin

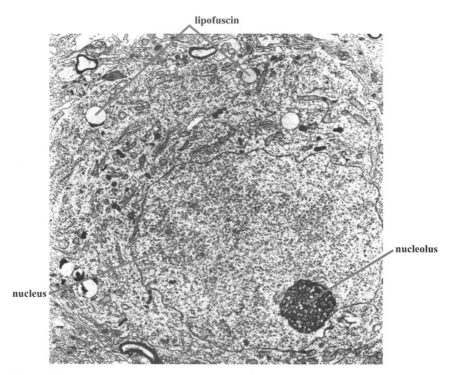

nucleolus

nucleus

Figure 2.1c   Neuronal soma with lipofuscin (pigment of age).

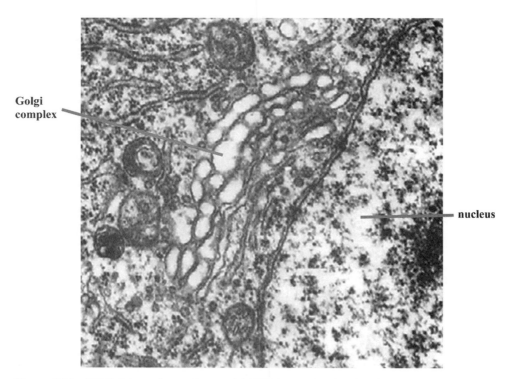

Golgi
complex

nucleus

Figure 2.1d   EM Golgi complex contiguous with RER in a perinuclear location.

axon terminal

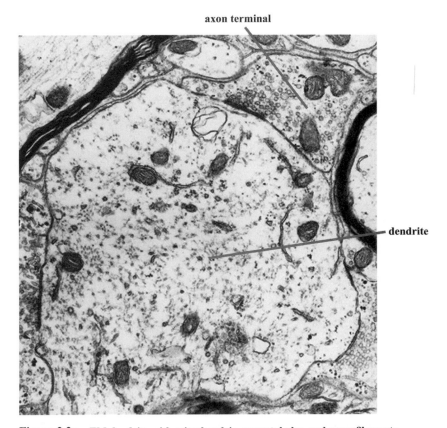

dendrite

Figure 2.2*a*   **EM dendrite with mitochondria, neurotubules, and neurofilaments.**

mitochondrion

neurofilaments

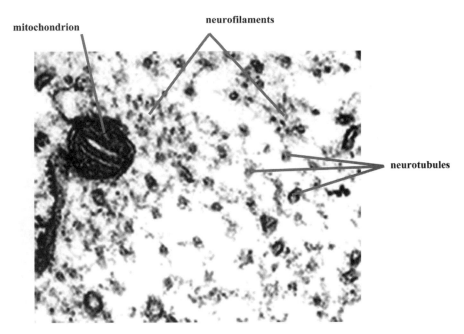

neurotubules

Figure 2.2*b*   **EM dendritic neurotubules and neurofilaments.**

dendrite

axospinous
terminal

mitochondria

dendritic
spine

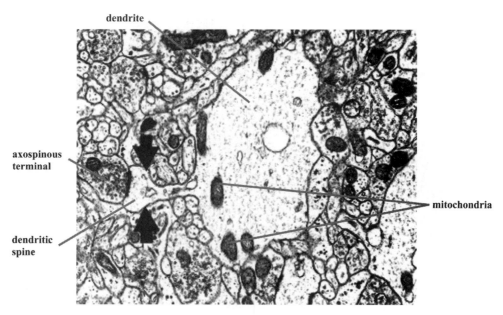

Figure 2.3*a*   EM dendrite with spine receiving axospinous terminal.

Axon terminals or **boutons** contain abundant **mitochondria** and **synaptic vesicles** and typically end in synapses that show **pre- and postsynaptic densities**. Where these densities are heavier on the postsynaptic membrane, the synapse is **asymmetric**. Where the densities are roughly equivalent, the synapse is **symmetric**. Typically asymmetric synapses contain predominantly **translucent** (clear) **round vesicles** that

axospinous synapse

spine apparatus

dendritic
spine

dendrite

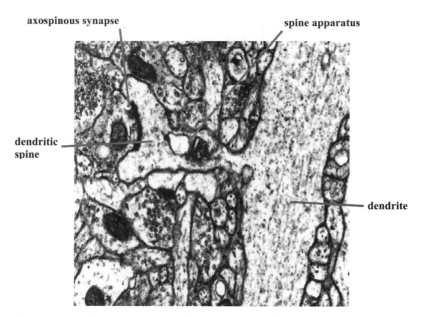

Figure 2.3*b*   EM dendritic spine with spine apparatus and smooth ER.

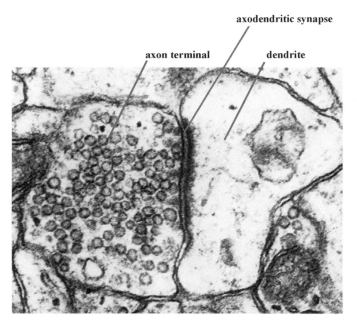

Figure 2.4a EM axodendritic synapse with clear round synaptic vesicles.

contain excitatory neurotransmitters (such as glutamate, aspartate) and are **excitatory** (Fig. 2.4a). Symmetric synapses contain predominantly **translucent** (clear) **flattened vesicles** that contain inhibitory neuro- transmitters [such as gamma aminobutyric acid (GABA), glycine] and are **inhibitory** (Fig. 2.4b). Axon terminals with round **dense**

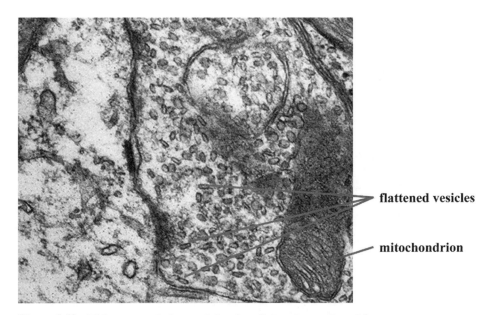

Figure 2.4b EM axon terminals containing clear flattened synaptic vesicles.

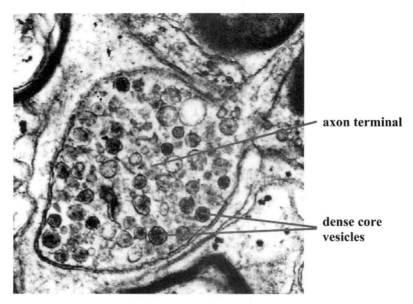

Figure 2.4c   EM axon terminal containing dense core synaptic vesicles.

core vesicles (Fig. 2.4c) that contain biogenic amines (such as norepinephrine, dopamine, serotonin) are usually **neuromodulatory** (metabotropic).

PROTOPLASMIC AND FIBROUS ASTROCYTES

At the EM level, **protoplasmic astrocytes** are located in gray matter between neuronal cell bodies and have organelle-rich cytoplasm with glycogen granules (Fig. 2.5a). **Fibrous astrocytes** found in white matter have a cytoplasm that typically contains abundant **glial filaments** (Fig. 2.5b).

INTERFASCICULAR OLIGODENDROCYTES

In the CNS the white matter contains large numbers of **myelinated axons** that are easily identifiable due to the black density of the concentric lamellae of the myelin sheath surrounding the axons (Fig. 2.6a). Occasionally an EM section fortuitously shows an **interfasicular oligodendrocyte**, the neuroglial cell that produces the myelin sheath. **Perineuronal oligodendrocytes** are close to the cell body where they myelinate the first segment of the axon (Fig. 2.6b).

SCHWANN-CELLS AND MYELINATED AND UNMYELINATED AXONS

In the PNS the peripheral nerves contain **Schwann cells** (neurilemmal cells) whose membranous extensions wrap the axon in concentric lamellae of **myelin** (Fig. 2.7a,b). The **major dense lines** of the myelin sheath are produced by the fusion of the inner

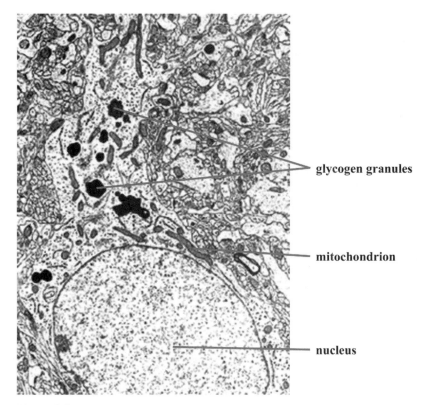

Figure 2.5*a*   EM protoplasmic astrocyte with glycogen granules.

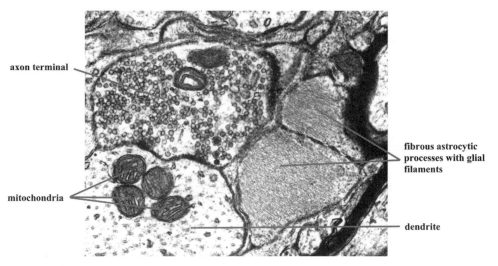

Figure 2.5*b*   EM fibrous astrocyctic process next to axodendritic synapse.

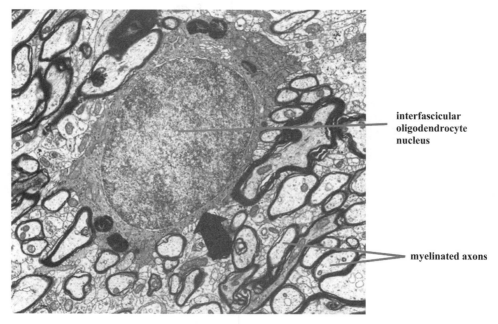

interfascicular
oligodendrocyte
nucleus

myelinated axons

Figure 2.6*a*   EM interfascicular oligodendrocyte in abundant myelinated axons.

perineuronal
oligodendrocyte
nucleus

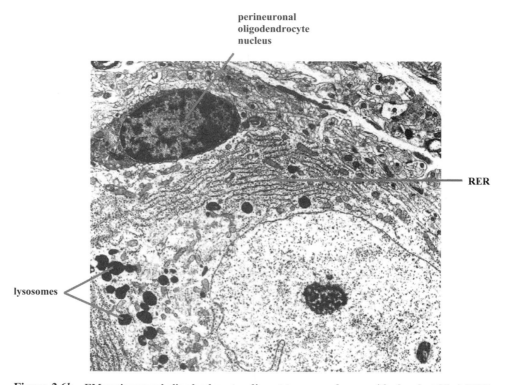

RER

lysosomes

Figure 2.6*b*   EM perineuronal oligodendrocyte adjacent to neuronal soma with abundant Nissl (REF)

axon

major dense
lines

myelin

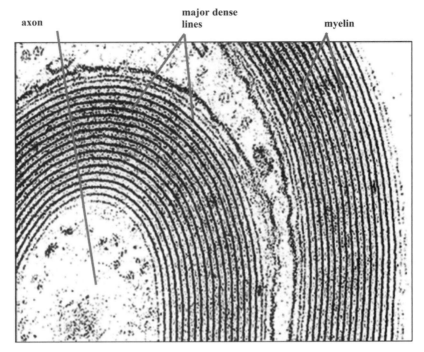

Figure 2.7a   Myelinated axons in peripheral nerve with Schwann cells.

myelinated
axons

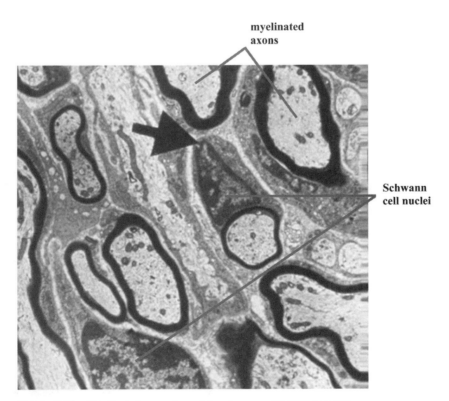

Schwann
cell nuclei

Figure 2.7b   Myelinated axons in peripheral nerve with Schwann cells.

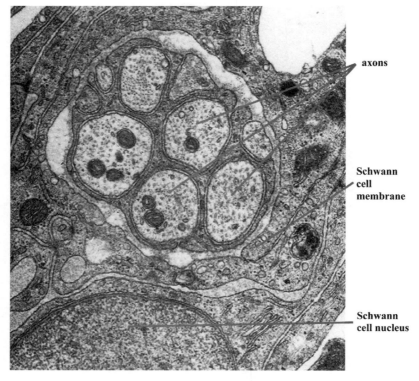

axons

Schwann cell membrane

Schwann cell nucleus

Figure 2.7c   **EM unmyelinated axons surrounded by Schwann cell membrane.**

(cytoplasmic) leaflets of the Schwann cell membrane, and the **intraperiod lines** are produced by the fusion of the outer surfaces of the Schwann cell membrane.

In some cases a Schwann cell membrane invests multiple small axons without generating individual concentric wrappings of myelin. These **unmyelinated nerves** (Fig. 2.7c), however, are still insulated from each other and typically correspond to slow-conducting C fibers (carry pain).

Chapter 3

Skull, Meninges, and Spinal Cord

ANTERIOR, MIDDLE, AND POSTERIOR CRANIAL FOSSAE AND CRANIAL FORAMINA

The interior of the cranial vault contains three depressions: the **anterior**, **middle**, and **posterior cranial fossae** (Fig. 3.1a). The anterior cranial fossa holds the inferior aspect

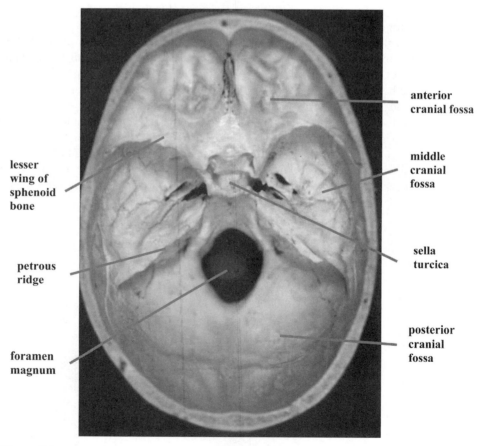

Figure 3.1a   Skull interior showing anterior, middle, and posterior cranial fossae.

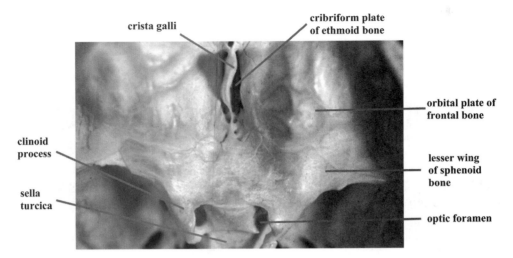

Figure 3.1*b* Anterior cranial fossa.

of the frontal lobes. The middle cranial fossa holds the poles of the temporal lobes. The posterior cranial fossa holds the cerebellum and brainstem.

On the midline of the anterior cranial fossa, the **cribriform plate of the ethmoid bone**, a perforated shelf of bone on either side of the **crista galli** (attachment for falx cerebri), transmits the **olfactory nerves (cranial nerve I)**. The orbital plate of the frontal bone forms the roof of the orbit. The **lesser wings of the sphenoid bone** occupy the caudal end of the fossa (Fig. 3.1b).

In the middle cranial fossa, the superior aspect of the sphenoid bone, between the lesser wings of the sphenoid, contains the **optic foramina**, which transmit the **optic nerves (CN II)**. The adjacent **anterior clinoid processes** serve as attachments for the tentorium cerebelli. The **sella turcica** is a depression in the superior midline of the body of the sphenoid in which the **pituitary gland** rests. Just under the lesser wings, the wide

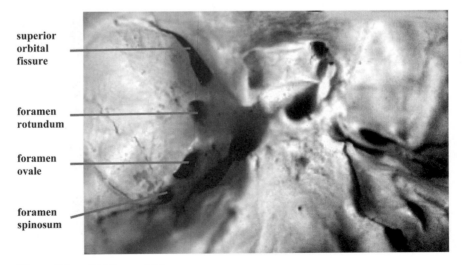

Figure 3.1*c* Middle cranial fossa.

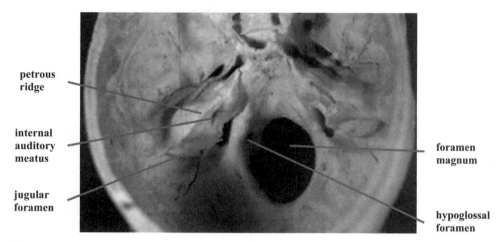

petrous
ridge

internal
auditory
meatus

jugular
foramen

foramen
magnum

hypoglossal
foramen

Figure 3.1d   **Posterior cranial fossa.**

superior orbital fissure transmits the **oculomotor (CN III), trochlear (CN IV), and abducens (CN VI) cranial nerves,** three cranial nerves that innervate extraocular muscles, into the orbit. On the floor of the fossa on the sides of the body of the sphenoid, the **foramen rotundum** and **foramen ovale** transmit the **maxillary** and **mandibular divisions of the trigeminal nerve (CN V),** respectively. The **foramen spinosum** transmits the middle meningeal artery, the principal blood supply of the dura mater (Fig. 3.1c).

The **petrous portion of the temporal bone** forms an elevated ridge that demarcates the middle from the posterior cranial fossa. Its **internal auditory meatus** transmits the **facial (CN VII)** and **vestibulocochlear (CN VIII) cranial nerves**. A channel-like depression for the **sigmoid sinus** in the lateral floor of the posterior cranial fossa ends into the **jugular foramen,** which transmits the **glossopharyngeal (CN IX), vagus (CN X),** and **spinal accessory (CN XI) cranial nerves** (Fig. 3.1d).

The **hypoglossal foramina,** which transmit the **hypoglossal nerves (CN XII),** are openings on either side of the large **foramen magnum** through which the caudalmost brainstem, the medulla, connects to the cervical spinal cord. The foramen magnum also transmits the **spinal accessory nerves (CN XI)** and **vertebral arteries.**

DURA MATER, DURAL REFLECTIONS, AND DURAL SINUSES

The **dura mater**, the toughest outer layer of the meninges, consists of two layers. An outer layer, the periosteal layer, forms the periosteum on the inside of the skull. **There is no epidural space in the skull**. The middle meningeal artery, the principal blood supply of the dura, courses within the periosteal layer and if torn the blood dissects between the bone and the dura (epidural hematoma) (Fig. 3.2a,b).

Its inner layer, the meningeal layer, gives rise to dural reflections (shelflike membranous partitions of dura) that extend into the brain. The **falx cerebri** extends into the longitudinal fissure, separating the two cerebral hemispheres (Fig. 3.3).

The **tentorium cerebelli** extends forward from the occiput to attach rostrally to the two petrous ridges and the **clinoid processes** of the sphenoid forming a roof over the posterior cranial fossa, and separating its contents (cerebellum and brainstem) from the inferior aspect of the occipital lobes (Fig. 3.4a). Rostrally, a large U-shaped opening,

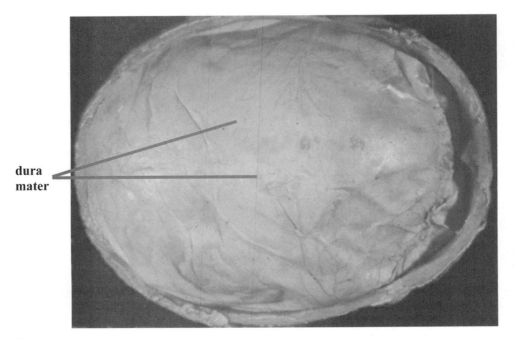

dura
mater

Figure 3.2a Craniectomy exposing dura mater.

the **tentorial notch** (or incisure) transmits the midbrain (rostralmost brainstem), connecting to the cerebrum in the "supratentorial compartment" of the skull (Fig. 3.4b). The **diaphragma sella** forms a roof over the sella turcica and has a small opening through which the pituitary stalk or infundibulum connects to the pituitary gland (Fig. 3.4a,b).

   Dural sinuses are endothelial-lined venous channels between the layers of the dura mater. The **superior sagittal sinus** lies in the attached margin of the falx cerebri. Most

dura
mater

arachnoid
membrane

Figure 3.2b Reflected dura mater reveals arachnoid membrane.

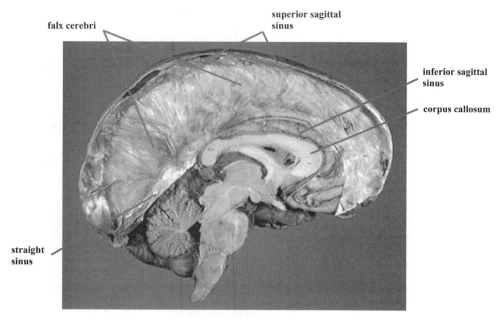

Figure 3.3 Falx cerebri is a major dural reflection between the cerebral hemispheres.

of the superficial **cerebral veins**, which run on the surface of the brain, traverse the subdural space (between arachnoid and dura) to drain into the superior sagittal sinus. The **inferior sagittal sinus** is in the free margin of the falx cerebri right above the corpus callosum. It is a tributary of the **straight sinus** in the attachment of the falx cerebri to the tentorium cerebelli and receives the **great cerebral vein of Galen**. The **transverse sinus** runs in the attached margin of the tentorium cerebelli and drains inferiorly into

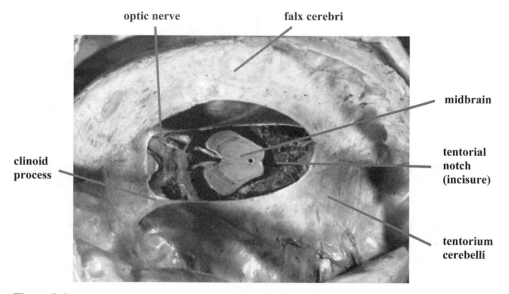

Figure 3.4a Tentorium cerebelli forms a roof over the posterior cranial fossa and has opening for brainstem.

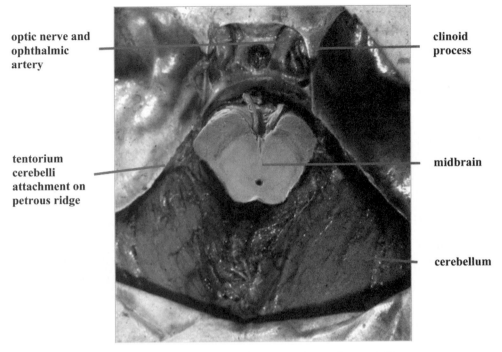

optic nerve and ophthalmic artery

clinoid process

tentorium cerebelli attachment on petrous ridge

midbrain

cerebellum

Figure 3.4*b* **Tentorium cerebelli removed reveals contents of posterior cranial fossa (cerebellum and brainstem).**

arachnoid granulations (villi)

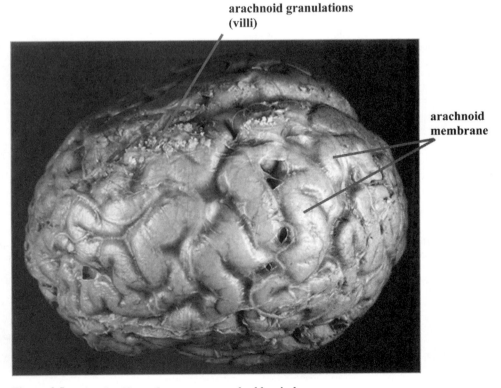

arachnoid membrane

Figure 3.5*a* **Arachnoid membrane covers cerebral hemisphere.**

the **sigmoid sinuses**, which ultimately drain into the **internal jugular vein** (exits the skull through the jugular foramen).

ARACHNOID MEMBRANE AND ARACHNOID GRANULATIONS

The opaque **arachnoid membrane** covers the surface of the brain but does not dip into sulci. In the living brain it "floats" above the cerebrospinal fluid-filled **subarachnoid space**. In its superiormost portion on either side of the midline, **arachnoid granulations** or **villi** are tufts of the membrane that protrude into the lateral lacunae of the superior sagittal sinus, and it is through these structures that cerebrospinal fluid (CSF) is reabsorbed into the systemic venous circulation (Fig. 3.5a,b). The arachnoid is connected to the superficial **pia mater** by filamentous **arachnoid trabeculae** that traverse the subarachnoid space.

The **pia mater** is the innermost meningeal layer that follows the contours of the brain. It consists of a superficial **epipial layer** that contains the blood vessels (brain arteries and veins), and an inner **intimal layer** that forms the pia-glial limiting membrane of the brain with astrocytic processes. Since the arteries run in the superficial pia on the floor of the subarachnoid space, if they are damaged there would be blood in the CSF.

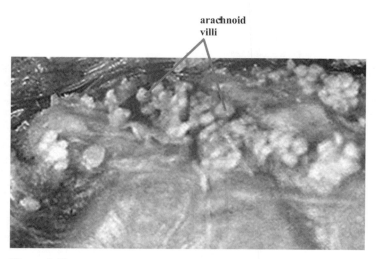

Figure 3.5*b*   **Arachnoid villi or granulations along dorsal margin protrude into superior sagittal sinus for reabsorption of CSF.**

GROSS ANATOMY OF THE SPINAL CORD

The spinal cord is about 18 inches in length, extending from the level of the foramen magnum to approximately the L2 vertebral level. The spinal cord is enlarged at those levels that contribute to the innervation of the upper and lower extremities: the **cervical** and **lumbar enlargements**, respectively. The spinal cord tapers into a cone-shaped ending, the **conus medullaris**, which ends at vertebral level L2. Below this level an extension of the pia mater descends as the **filum terminale**.

At vertebral level S2 (the bottom of the **dural sac**) the filum terminale is surrounded with dura, then called the **coccygeal ligament,** which anchors the cord inferiorly to the coccyx. The spinal cord is anchored laterally by about 20 pairs of **denticulate ligaments** (see Fig. 3.8a), toothlike connective tissue extensions where the pia pierces the arachnoid

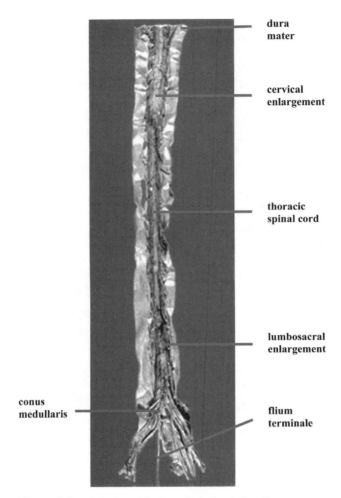

dura
mater

cervical
enlargement

thoracic
spinal cord

lumbosacral
enlargement

conus
medullaris

flium
terminale

Figure 3.6*a* **Spinal cord is about 18 inches in length.**

to attach to the dura. They come off the cord horizontally from the midpoint of the lateral funiculus, providing the neurosurgeon with a landmark separating the posterolateral from the anterolateral quadrant (Fig. 3.6 a,b).

There are 31 pairs of **spinal nerves**: 8 cervical, 12 thoracic, 5 lumbar, 5 sacral, and 1 coccygeal, and 31 segments of the cord, each segment giving rise to a pair of spinal nerves. The dorsal and ventral roots come together to form the spinal nerve at the level of the inter-vertebral foramen of exit. Spinal nerves C1 through C7 exit above the vertebra of the same number. C8 exits below vertebra C7. Then spinal nerves T1 through coccygeal-1 exit below the vertebrae of the same number.

The spinal nerves in the cervical region exit nearly horizontally through intervertebral foramina at about the same level, but beginning in the thoracic region the dorsal and ventral roots descend to their foramen of exit. Since the spinal cord ends at vertebral level L2, dorsal and ventral roots from lower levels are significantly lengthened to reach their level of exit, forming an aggregation of rootlets that resembles a horse's tail, the **cauda equina** (see Fig. 3.9a).

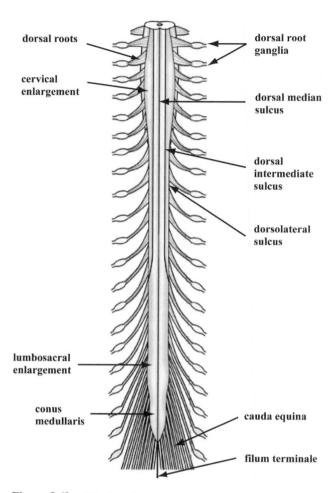

Figure 3.6*b* **Spinal cord gross anatomy.**

In the living cord CSF pressure in the subarachnoid space forces the arachnoid membrane against the inside of the dural sac so that it contains a large pool of CSF (**lumbar cistern**), which can be tapped in lumbar puncture. In this procedure, there is no spinal cord below vertebral level L2 that might potentially be damaged, and the nerve roots of the cauda equina in the dural sac deflect away from the needle.

The relative amount of white matter and gray matter in a cross section varies between levels of the spinal cord. In the cervical region the white matter is the greatest as its constitute tracts are at their largest. Sensory tracts get larger as one ascends the cord as contributions from progressively higher levels of the body are added. Motor tracts become smaller as one descends through the cord as projections are given off into the gray matter at progressively lower levels.

CERVICAL SPINAL CORD

In cross section the **cervical spinal cord** is ovoid and has the largest amount of white matter, the **dorsal**, **lateral**, and **ventral funiculi**. The **dorsal median sulcus**, extending

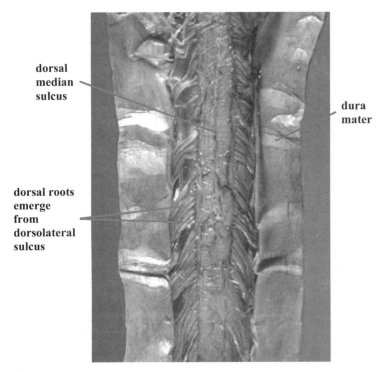

dorsal
median
sulcus

dura
mater

dorsal roots
emerge
from
dorsolateral
sulcus

Figure 3.7a   Cervical spinal cord with dura reflected.

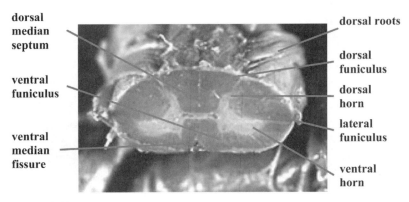

dorsal
median
septum

dorsal roots

dorsal
funiculus

ventral
funiculus

dorsal
horn

ventral
median
fissure

lateral
funiculus

ventral
horn

Figure 3.7b   Cervical spinal cord cross section.

into the cord as the **dorsal median septum**, divides the two dorsal funiculi (dorsal columns). In the cervical and upper thoracic regions the **dorsal funiculus** is divided by the **dorsal intermediate sulcus** into two major tracts: **fasciculus gracilis** and **cuneatus**. The **dorsal roots** emerge from the **dorsolateral sulcus**, whereas the **ventral roots** emerge from the **ventrolateral sulcus**. The deep **ventral median fissure** contains the **anterior spinal artery** (Fig. 3.7a). The **dorsal horn, intermediate zone**, and **ventral**

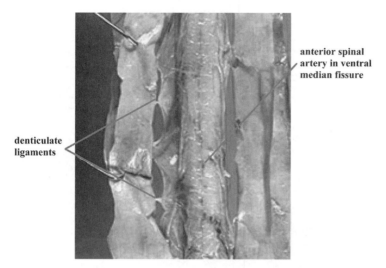

anterior spinal
artery in ventral
median fissure

denticulate
ligaments

Figure 3.8*a*   Thoracic spinal cord with dura reflected showing denticulate ligaments.

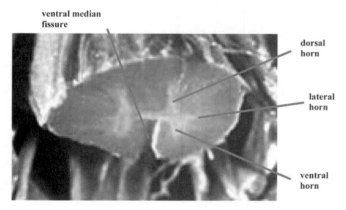

ventral median
fissure

dorsal
horn

lateral
horn

ventral
horn

Figure 3.8*b*   Thoracic spinal cord cross section.

horn of the spinal cord gray matter are enlarged because of the increased number of neurons involved with the innervation of the upper extremity (Fig. 3.7b).

THORACIC SPINAL CORD

In cross section the **thoracic spinal cord** is round and has the smallest amount of gray matter, which has an H-shaped configuration. In the upper thoracic cord the dorsal funiculus is divided by the dorsal intermediate sulcus into two tracts, but in the lower thoracic cord the sulcus is not present and the dorsal funiculus only contains the fasciculus gracilis. The **lateral horn** of the spinal cord gray matter is only present from T1 to L2 and thus is a prominent characteristic of the thoracic cord. It contains the intermediolateral nucleus (preganglionic sympathetic neurons).

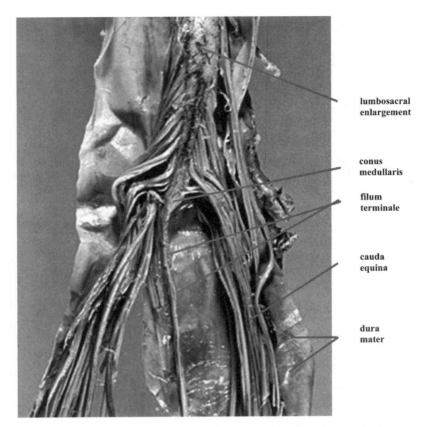

lumbosacral
enlargement

conus
medullaris

filum
terminale

cauda
equina

dura
mater

Figure 3.9*a*   Lumbosacral spinal cord with cauda equina and filum terminale.

LUMBAR AND SACRAL SPINAL CORD

In cross section the **lumbar spinal cord** is round but larger in diameter than the thoracic cord because the lumbosacral enlargement is related to the innervation of the lower

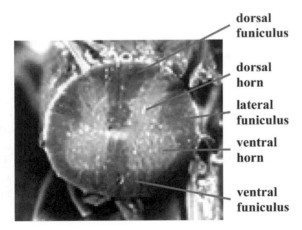

dorsal
funiculus

dorsal
horn

lateral
funiculus

ventral
horn

ventral
funiculus

Figure 3.9*b*   Lumbar spinal cord cross section.

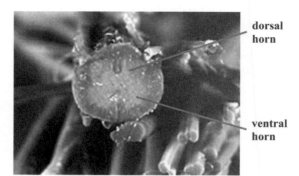

dorsal
horn

ventral
horn

Figure 3.9c   Sacral spinal cord cross section.

extremity. The ventral horns are enlarged because they contain motor neurons to muscles of the leg. The amount of white matter is proportionately smaller than in higher regions of the cord because the sensory and motor tracts are diminished in size (Fig. 3.9b).

   In cross section the **sacral spinal cord** (S1–S3) still contributes to the lumbosacral plexus and the innervation of the lower extremity and therefore has a significant amount of gray matter, but the white matter is minimal because the sensory and motor tracts are at their smallest (Fig. 3.9c).

Chapter 4

Gross Anatomy of the Brain

CEREBRUM: SUBDIVISION INTO LOBES

The **cerebrum** has five lobes: **frontal, parietal, occipital, temporal,** and **insular.** In a lateral view of the brain, the **central sulcus** runs obliquely downward and forward from approximately the midpoint of the dorsal margin of the cerebral hemisphere, ending just before it reaches the lateral sulcus. It separates the **frontal lobe** from the **parietal lobe** (Fig. 4.1a,b) The **lateral sulcus** runs horizontally on the lateral surface separating the frontal and parietal lobes from the **temporal lobe.** There is no prominent sulcus delineating the parietal and temporal lobes from the **occipital lobe,** so an imaginary line is drawn from the **preoccipital notch,** an indentation on the inferior margin of the hemisphere, to the point where the parietooccipital sulcus crosses the dorsal midline.

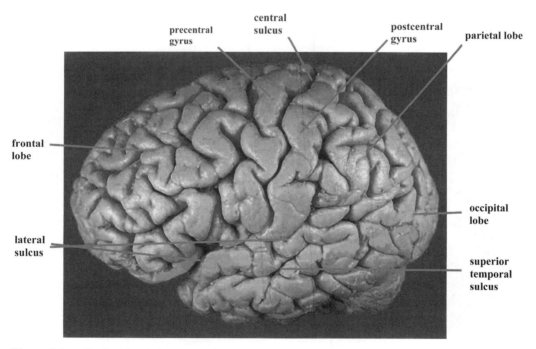

Figure 4.1a   Cerebral hemisphere, lateral view.

Digital Neuroanatomy, by George R. Leichnetz
Copyright © 2006 John Wiley & Sons, Inc.

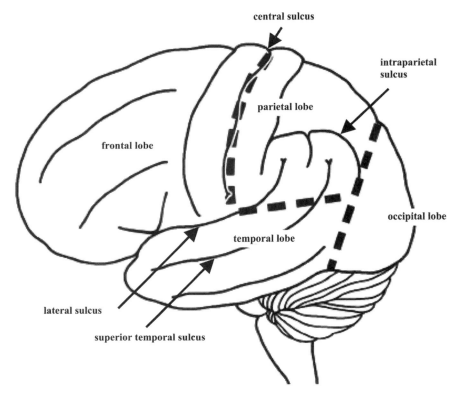

Figure 4.1*b* Cerebral lobes.

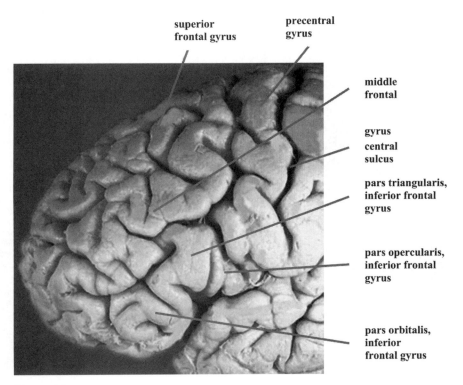

Figure 4.2 Frontal lobe.

FRONTAL LOBE

The **precentral gyrus**, the primary motor cortex, is the vertically running gyrus immediately in front of the central sulcus. It is the caudalmost gyrus of the frontal lobe. The remainder of the frontal lobe is made up of three gyri running horizontally perpendicular to the precentral gyrus, the **superior**, **middle**, and **inferior frontal gyri** separated by the **superior** and **inferior frontal sulci** (Fig. 4.2). The inferior frontal gyrus is subdivided into three parts: **pars opercularis**, **pars triangularis**, and **pars orbitalis**. In the left dominant hemisphere, the former two constitute **Broca's motor speech area**.

PARIETAL LOBE

The **postcentral gyrus**, the primary somatosensory cortex, is the vertically running gyrus immediately behind the central sulcus. It is the rostralmost gyrus of the parietal lobe. The remainder of the parietal lobe is made up by two lobules, the **superior** and **inferior parietal lobules,** separated by the **intraparietal sulcus** that runs horizontally perpendicular to the postcentral gyrus (Fig. 4.3). The inferior parietal lobule is further subdivided into two gyri: the **supramarginal gyrus**, surrounding the caudal end of the lateral sulcus, and the **angular gyrus**, surrounding

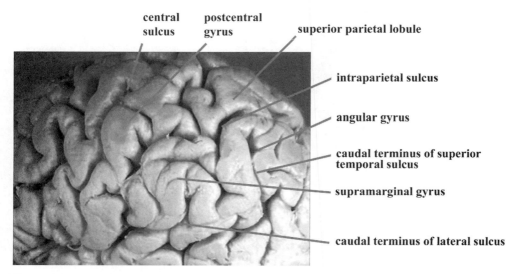

Figure 4.3   Parietal lobe.

the caudal end of the superior temporal sulcus. In the left dominant hemisphere, these two gyri make up part of **Wernicke's area**, a cortical area concerned with language comprehension.

INSULAR LOBE AND TEMPORAL LOBE

The **insular lobe** is hidden in the depths of the lateral sulcus (Fig. 4.4). The **superior transverse temporal gyri**, the primary auditory cortex, are two gyri that run obliquely across the superior aspect of the temporal lobe. The lateral aspect of the **temporal lobe**

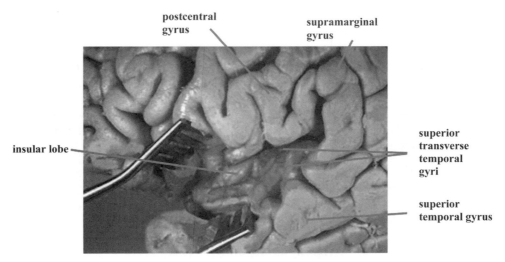

Figure 4.4   Insular lobe, deep within lateral sulcus.

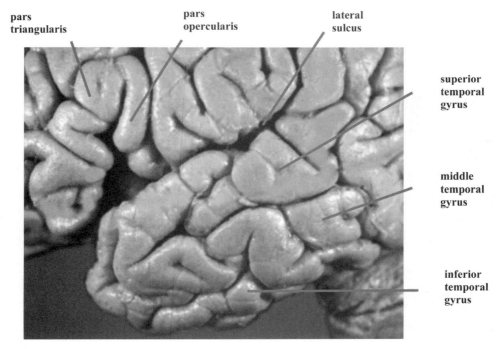

Figure 4.5   Temporal lobe.

shows three gyri parallel to the lateral sulcus, the **superior**, **middle**, and **inferior temporal gyri** separated by the **superior** and **middle temporal sulci** (Fig. 4.5).

OCCIPITAL LOBE

The lateral aspect of the **occipital lobe**, which lies caudal to the imaginary line drawn from the preoccipital notch to the dorsal margin of the hemisphere, shows **lateral occipital gyri**.

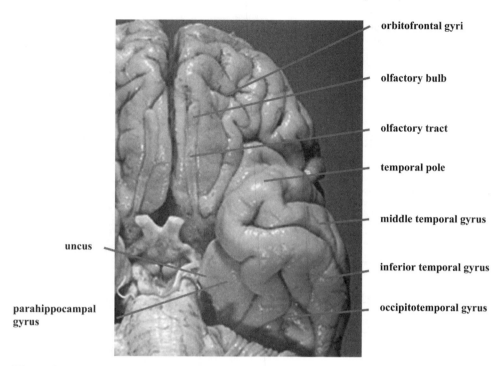

orbitofrontal gyri

olfactory bulb

olfactory tract

temporal pole

middle temporal gyrus

uncus

inferior temporal gyrus

parahippocampal gyrus

occipitotemporal gyrus

Figure 4.6 Inferior aspect of frontal and temporal lobes.

INFERIOR ASPECT OF FRONTAL AND TEMPORAL LOBES

The inferior aspect of the frontal lobe contains the **orbitofrontal gyri,** and the **olfactory bulb** and **tract** (associated with cranial nerve I) (Fig. 4.6). The inferior aspect of the temporal lobe from lateral to medial shows the **inferior temporal gyrus**, **inferior temporal sulcus, occipitotemporal gyrus, collateral sulcus**, and **parahippocampal gyrus**. The **uncus** is an elevation on the medial aspect of the parahippocampal gyrus that overlies the amygdala and rostral part of the hippocampus. This gyrus lies along the margin of the tentorial notch such that with an expanding mass in the supratentorial compartment the uncus herniates through the opening (uncal herniation) compressing the midbrain.

DIENCEPHALON AND MIDBRAIN

The ventral aspect of the diencephalon contains the **optic nerves (CN II)**, **optic chiasm**, and **optic tracts**, as well as **tuber cinereum** (elevation on the ventral aspect of the hypothalamus) and **mammillary bodies** (Fig. 4.7).

The ventral aspect of the **midbrain** contains the **cerebral peduncles** and the depression between them, the **interpeduncular fossa**, from which the **oculomotor nerves (CN III)** emerge. The **trochlear nerves (CN IV)**, which exit from the dorsal aspect of the midbrain, course ventrally around the cerebral peduncles to join the oculomotor nerves en route to the orbit where they supply extraocular muscles.

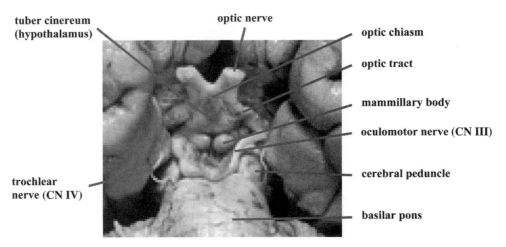

tuber cinereum (hypothalamus)

optic nerve

optic chiasm

optic tract

mammillary body

oculomotor nerve (CN III)

cerebral peduncle

basilar pons

trochlear nerve (CN IV)

Figure 4.7 Ventral aspect of diencephalon and midbrain.

PONS AND MEDULLA

The ventral aspect of the **pons** contains the **basilar pons** and the large bundles that connect it to the cerebrellum, the **middle cerebellar peduncles**. The large **trigeminal nerves (CN V)** come off the lateral part of the basilar pons.

The **pontomedullary junction** is a line that delineates the basilar pons from the **medulla**. The **abducens nerves (CN VI)** exit the pontomedullary junction in line with the preolivary sulcus. More laterally the **facial (CN VII)** and **vestibulocochlear (CN VIII) cranial nerves** emerge from the **cerebellopontine angle** next to the **flocculus** of the cerebellum (Fig. 4.8).

On the ventral aspect of the medulla the **medullary pyramids** (containing the pyramidal tracts) run vertically on either side of the **ventral median fissure**, which is interrupted by the crossing of the tracts in the **pyramidal decussation**. The pyramids are separated from the **olives** by the **preolivary sulcus** from which the roots of the **hypoglossal nerves (CN XII)** emerge. The **glossopharyngeal (CN IX)** and **vagus (CN X) nerves** exit from the postolivary sulcus.

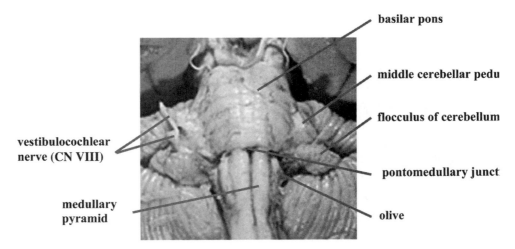

basilar pons

middle cerebellar pedu

flocculus of cerebellum

pontomedullary junct

olive

vestibulocochlear nerve (CN VIII)

medullary pyramid

Figure 4.8 Ventral aspect of pons and medulla.

CEREBELLUM AND CEREBELLAR PEDUNCLES

The **cerebellum** has three lobes: **anterior lobe, posterior lobe,** and **flocculonodular lobe.** Each of the lobes has a midline portion in the **vermis** and a lateral portion, **hemisphere.** The **primary fissure** separates the anterior and posterior lobes (Fig. 4.9a). The **prenodular fissure** (posterolateral fissure in hemisphere) separates the posterior lobe from the flocculonodular lobe. The cerebellar vermis has 10 sublobules (Fig. 4.9b,c).

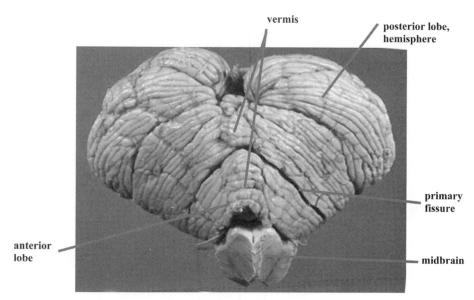

Figure 4.9a   Superior aspect of cerebellum with midbrain.

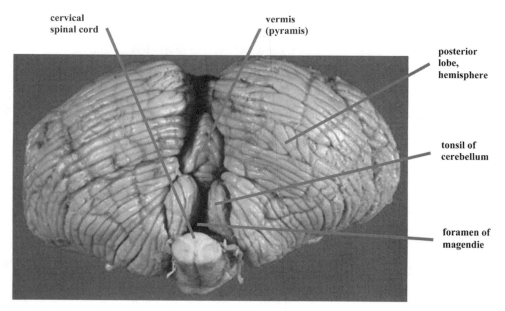

Figure 4.9b   Inferior aspect of cerebellum with cervical spinal cord.

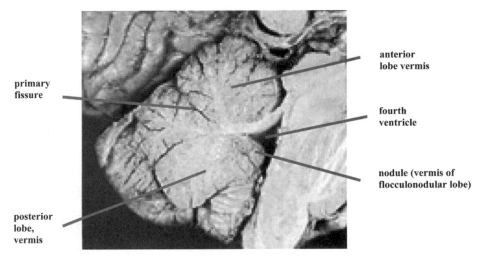

primary
fissure

anterior
lobe vermis

fourth
ventricle

nodule (vermis of
flocculonodular lobe)

posterior
lobe,
vermis

Figure 4.9c   Midsagittal view of cerebellar vermis.

The cerebellar peduncles are large bundles that connect the cerebellum to the brainstem and carry cerebellar tracts. The **superior cerebellar peduncle** connects the cerebellum to the midbrain. The **middle cerebellar peduncle** connects the cerebellum to the pons (Fig. 4.10). The **inferior cerebellar peduncle** connects the cerebellum to the medulla.

superior cerebellar
peduncle

midbrain

middle
cerebellar
peduncle

Figure 4.10   Cerebellar peduncles with anterior lobe hemisphere removed.

BRAINSTEM (CEREBRAL CORTEX AND CEREBELLUM REMOVED)

Removing the cerebral cortex and subcortical white matter, and the cerebellum exposes the dorsal aspect of the **thalamus** and **brainstem** (Figs. 4.11, 4.12).

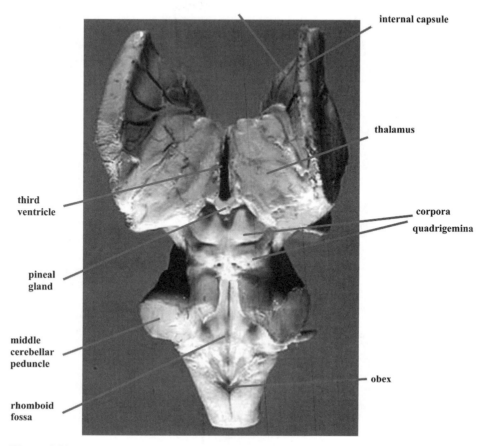

Figure 4.11   Dorsal view of basal ganglia, thalamus, and brainstem with cerebral cortex and cerebellum removed.

DIENCEPHALON (THALAMUS, PINEAL GLAND)

The two thalami lie on either side of the midline **third ventricle** (Fig. 4.13). The **stria terminalis** (and terminal vein) run in a groove in the floor of the lateral ventricle and delineate the thalamus from the **caudate nucleus**. The **internal capsule** separates the caudate nucleus from the **putamen**. The **pineal gland** and **habenula** are part of the epithalamus that come off the caudal superior end of the third ventricle.

MESENCEPHALON (MIDBRAIN)

The dorsal aspect of the **midbrain** contains four elevations, the **corpora quadrigemina**, which include the paired **superior** and **inferior colliculi** (Fig. 4.14). The **trochlear nerves (CN IV)** exit from tiny openings in the anterior medullary velum just behind the inferior colliculi.

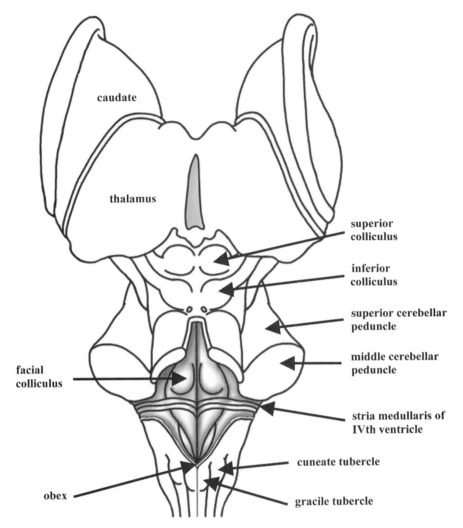

Figure 4.12   Dorsal aspect of thalamus and brainstem.

PONS AND MEDULLA, RHOMBOID FOSSA

The **anterior medullary velum** is a thick membrane that connects the two superior cerebellar peduncles, forming a roof over the rostral part of the **fourth ventricle**. The floor of the fourth ventricle is the diamond-shaped **rhomboid fossa** (Fig. 4.15). Striations (**stria medullaris of the fourth ventricle**) run horizontally dividing the rhomboid fossa into two triangles: a rostral triangle over the pons and a caudal triangle over the medulla. The **dorsal median sulcus** is a midline groove running rostrocaudally in the rhomboid fossa, caudally contiguous with the same sulcus in the dorsal aspect of the spinal cord. The **sulcus limitans** is a groove somewhat more lateral, a remnant of the embryonic sulcus limitans, separating medial motor nuclei from sensory nuclei on the floor of the fossa. The **facial colliculi** are elevations that lie in the caudal dorsal pons between the dorsal median sulcus and sulcus limitans, produced by the facial nerve coursing over the abducens nucleus. The **obex** is the caudalmost point of the rhomboid fossa.

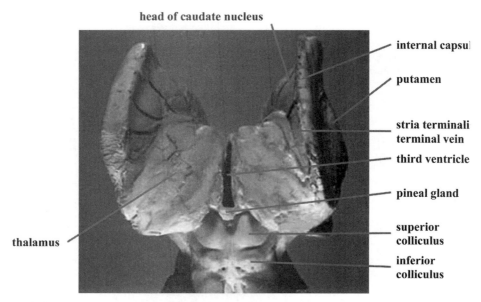

head of caudate nucleus

internal capsul

putamen

stria terminali
terminal vein

third ventricle

pineal gland

superior
colliculus

inferior
colliculus

thalamus

Figure 4.13   Dorsal aspect of thalamus and basal ganglia with midbrain.

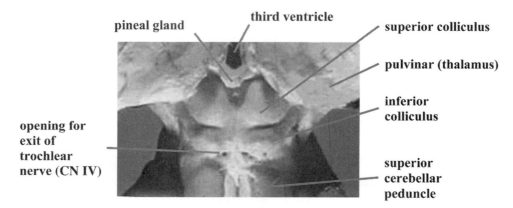

pineal gland

third ventricle

superior colliculus

pulvinar (thalamus)

inferior
colliculus

opening for
exit of
trochlear
nerve (CN IV)

superior
cerebellar
peduncle

Figure 4.14   Dorsal aspect of midbrain.

The dorsal aspect of the **medulla** contains the **gracile tubercle** (or clava) overlying the nucleus gracilis and the **cuneate tubercle** overlying the cuneate nucleus (Fig. 4.16). The **fasciculus gracilis** is a tract that continues rostrally from the dorsal columns of the spinal cord to terminate in the nucleus gracilis. The **fasciculus cuneatus**, separated from the fasciculus gracilis by the dorsal intermediate sulcus, terminates in the nucleus cuneatus.

CRANIAL NERVES

With the superficial pia mater and blood vessels intact, the **cranial nerves** are visualized more naturally, including the **olfactory tract** (associated with CN I), **optic nerve** (CN II) and **optic chiasm** with adjacent internal carotid artery immediately rostral to the tuber cinereum of the hypothalamus, the **oculomotor nerve** (CN III) emerging from the inter-peduncular fossa between the posterior cerebral and superior cerebellar arteries, the large

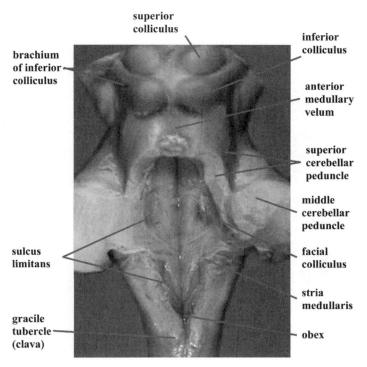

Figure 4.15 Dorsal aspect of brainstem with cerebellum removed, showing floor of fourth ventricle (rhomboid fossa) above pons and medulla.

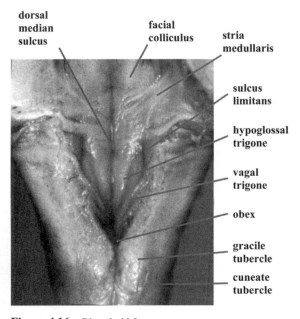

Figure 4.16 Rhomboid fossa.

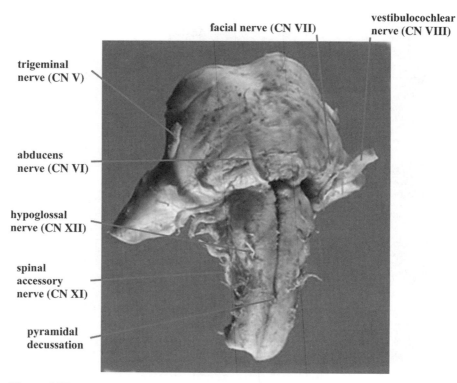

Figure 4.17a   Brainstem with cranial nerves.

trigeminal nerve (CN V) exiting the middle cerebellar peduncle, the **facial** (CN VII), **vestibulocochlear** (CN VIII) **nerves** in the cerebellopontine angle, **vagus** (CN X), and **glossopharyngeal** (CN IX) **nerves** exiting the postolivary sulcus, and the **hypoglossal nerves** (CN XII) exiting the preolivary sulcus.

MIDSAGITTAL SECTION (MEDIAL ASPECT OF CEREBRAL HEMISPHERE)

The medial aspect of the brain is studied best on a midsagittal section. The frontal lobe extends caudally to the point where the central sulcus notches the midline where the **superior frontal gyrus** meets the **paracentral lobule** (Figs. 4.18,4.19). The rostral half of the paracentral lobule is a continuation of the primary motor cortex (precentral gyrus) and the caudal half is a continuation of the primary somatosensory cortex (postcentral gyrus). Both have a representation of the leg. The **limbic lobe** consists in part of frontal, parietal, and temporal lobes. The **subcallosal gyrus** lies under the genu of the corpus callosum and is contiguous with the **cingulate gyrus** above the corpus callosum, which extends inferiorly into the **parahippocampal gyrus**. The parietal lobe is separated from the occipital lobe by the **parietooccipital sulcus**. The medial aspect of the occipital lobe contains the **cuneus** and **lingual gyri**, the primary visual cortex, separated by the **calcarine fissure**.

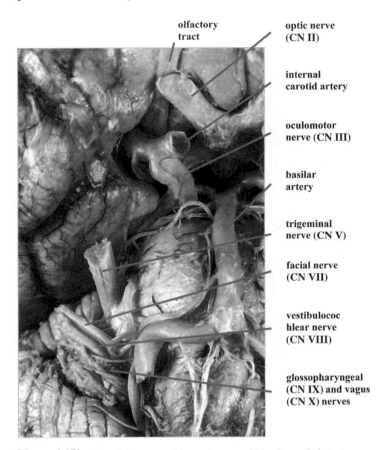

Figure 4.17*b*   Cranial nerves with meninges and blood vessels intact.

The **corpus callosum** is a bridge between the two cerebral hemispheres, the largest commissure of the brain, interconnecting homologous cortical areas. It has a **genu**, **body**, and **splenium**. The **septum pellucidum** is a vertical membranous partition that extends from the corpus callosum to the **fornix**, separating the two **lateral ventricles** (the cavities within the hemispheres). The fornix is a large tract that carries efferents from the hippocampus.

The **interventricular foramen of Monro** connects each of the lateral ventricles to the midline **third ventricle** that separates the two thalami (Fig. 4.20). The **massa intermedia** or **interthalamic adhesion** is an adhesion of ependymal and glial cells across the third ventricle. The **pineal gland** and **habenula** belong to the epithalamus and are connected to the caudal superior aspect of the diencephalon. The **hypothalamus** is the ventralmost diencephalon that is delineated from the thalamus by the **hypothalamic sulcus**, a line in the wall of the third ventricle. The **anterior commissure** interconnects the temporal lobes. The **lamina terminalis** is a vertical membrane that runs from the anterior commissure to the **optic chiasm** and embryologically closed the rostral end of the developing neural tube, the anterior neuropore. The **pituitary stalk** or **infundibulum** connects the hypothalamus to the pituitary gland. The **mammillary bodies** are also part of the hypothalamus.

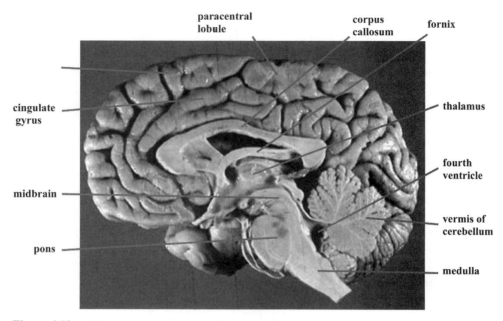

Figure 4.18   Midsagittal aspect of the cerebral hemisphere.

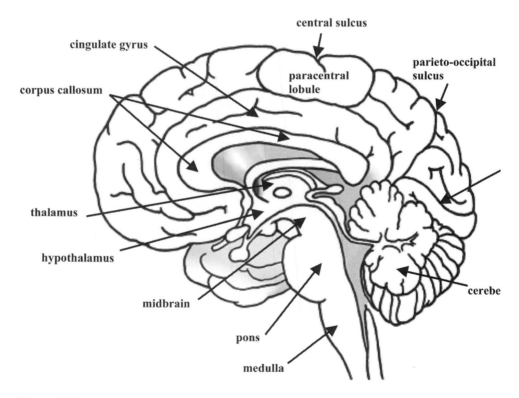

Figure 4.19   Midsagittal section of the brain.

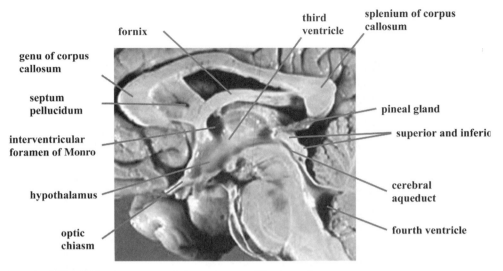

genu of corpus callosum

septum pellucidum

interventricular foramen of Monro

hypothalamus

optic chiasm

fornix

third ventricle

splenium of corpus callosum

pineal gland

superior and inferior

cerebral aqueduct

fourth ventricle

Figure 4.20   Midsagittal aspect of diencephalon, midbrain, and pons.

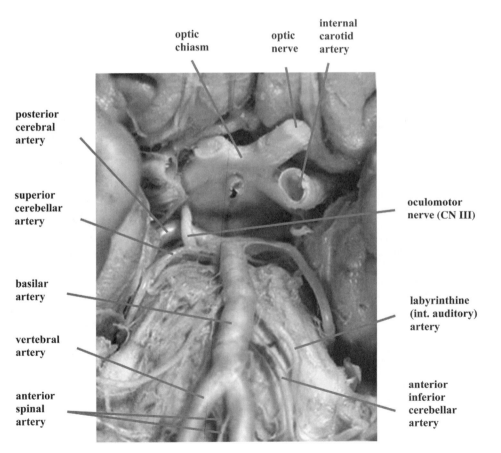

optic chiasm

optic nerve

internal carotid artery

posterior cerebral artery

superior cerebellar artery

basilar artery

vertebral artery

anterior spinal artery

oculomotor nerve (CN III)

labyrinthine (int. auditory) artery

anterior inferior cerebellar artery

Figure 4.21   Circle of Willis and brainstem arteries.

The **cerebral aqueduct** of the midbrain connects the midline third ventricle to the fourth ventricle. The region above the cerebral aqueduct is referred to as the midbrain **tectum** and contains the corpora quadrigiemina (**superior** and **inferior colliculi**). The region below the aqueduct is the **midbrain tegmentum**.

The midline **vermis** of the cerebellum forms the roof of the fourth ventricle. The floor of the fourth ventricle is the **rhomboid fossa**, which overlies the pons and medulla. The pons is divided into the **pontine tegmentum** and **basilar pons**.

BLOOD SUPPLY OF THE BRAIN

The blood supply of the brain comes through two major arterial systems: the internal carotid system and vertebrobasilar system that anastamose on the base of the brain in the **cerebral arterial circle of Willis**. The **internal carotid artery** enters the skull through the internal carotid foramen, joins the circle of Willis, and gives rise to the anterior and middle cerebral arteries. The **vertebral arteries** enter the skull through the foramen magnum and join to form the **basilar artery**. Before joining, the vertebrals give rise to the paired **posterior spinal arteries** and unpaired **anterior spinal artery,** and the **posterior inferior cerebellar arteries**. The basilar artery gives off the **anterior inferior cerebellar arteries, labyrhinthine** (or internal auditory) **arteries, pontine arteries,** and **superior cerebellar arteries**, before bifurcating into its terminal branches, the **posterior cerebral arteries** (Figs. 4.21,4.22).

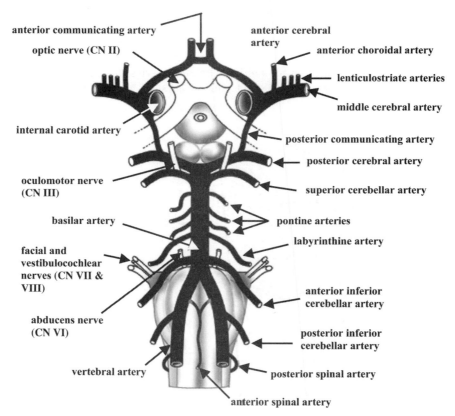

Figure 4.22 Circle of Willis and brainstem arteries.

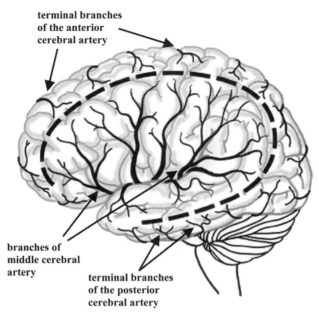

terminal branches
of the anterior
cerebral artery

branches of
middle cerebral
artery

terminal branches
of the posterior
cerebral artery

Figure 4.23*a*   Middle cerebral arterial distribution.

The **anterior cerebral** arteries, which are derived from the internal carotid arteries, are connected through the **anterior communicating artery**. The branches of the anterior cerebral arteries supply the medial aspect of the frontal and parietal lobes through the parietooccipital sulcus (Figs. 4.23a,b). The **middle cerebral arteries** originate from the internal carotid artery and course laterally, giving off the small caliber **anterior choroidal arteries** and multiple **lenticulostriate arteries** that perforate the basal fore-brain to supply the basal ganglia and internal capsule. The main trunk of the middle

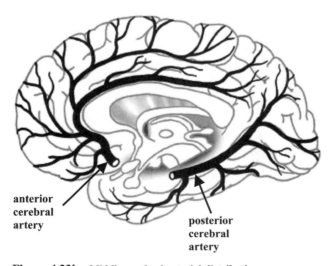

anterior
cerebral
artery

posterior
cerebral
artery

Figure 4.23*b*   Middle cerebral arterial distribution.

cerebral artery emerges from the lateral sulcus and its branches supply the lateral aspect of the cerebral hemisphere. The **posterior cerebral arteries** are derived from the bifurcation of the basilar artery. After giving off branches to the basal midbrain, they supply the inferior aspect of the temporal lobe and medial aspect of the occipital lobe.

Chapter 5

Sectional Anatomy of the Brain

CORONAL SECTION: HEAD OF CAUDATE NUCLEUS

In a coronal section through the rostral frontal lobe the **superior, middle**, and **inferior frontal gyri** are separated by the **superior** and **inferior frontal sulci** (Fig. 5.1). The **orbitofrontal gyri** are on the inferior aspect of the section. The **head of the caudate nucleus** is located in the lateral wall of the **anterior horn of the lateral ventricle.** The **genu of the corpus callosum** contains commissural fibers interconnecting homolgous areas of the frontal cortex. The **septum pellucidum** separates the lateral ventricles.

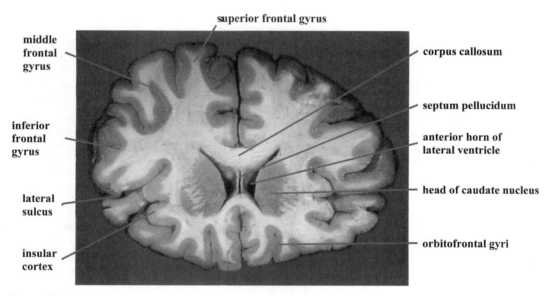

Figure 5.1 Coronal section through frontal lobe, head of caudate nucleus.

CORONAL SECTION: STRIATUM

The **anterior limb of the internal capsule** separates the head of the caudate nucleus from the putamen, known collectively as the **striatum** (Fig. 5.2). The **cingulate gyrus** (part of the limbic lobe) lies above the corpus callosum.

Digital Neuroanatomy, by George R. Leichnetz
Copyright © 2006 John Wiley & Sons, Inc.

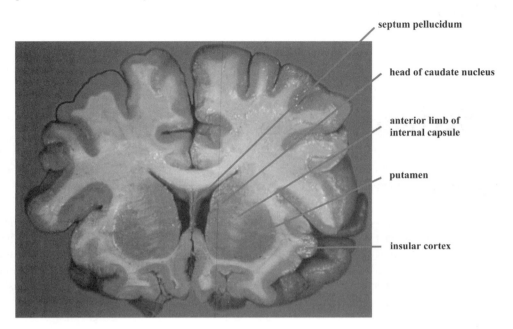

Figure 5.2   Coronal section through frontal lobe, showing striatum.

CORONAL SECTION: ANTERIOR COMMISSURE

The **anterior commissure** interconnects homologous areas of the lower portions of the temporal lobe. Although visualized best in horizontal section, a coronal section at the level of the anterior commissure passes through the **genu of the internal capsule.** The **putamen** is delineated from the **globus pallidus** by a lamina, and the two structures are known collectively as the **lentiform nucleus** (Fig. 5.3). The **external** capsule separates

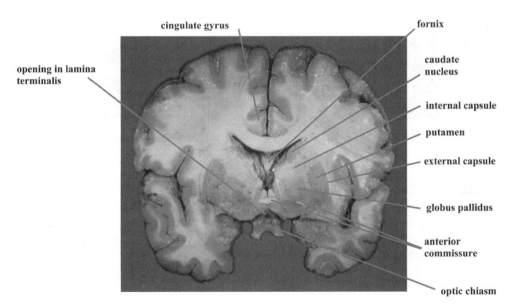

Figure 5.3   Coronal section at level of anterior commissure.

the lentiform nucleus from the claustrum, and the **extreme capsule** separates the claustrum from the **insular** cortex. The **lamina terminalis** extends from the anterior commissure to the **optic chiasm** and is a membrane that closes the rostral end of the third ventricle (which was the anterior neuropore of the embryonic neural tube).

CORONAL SECTION: OPTIC CHIASM, OPTIC TRACTS, AMYGDALA

The **thalamus** occupies the walls of the dorsal part of the **third ventricle** (Fig. 5.4). The **septum pellucidum** is a vertical partition between the lateral ventricles that runs from the **corpus callosum** to the **fornix.** At this level, the **interventricular foramen of Monro** connects the lateral ventricle to the third ventricle. The **posterior limb of the internal capsule** separates the thalamus from the lentiform nucleus. The **hypothalamus** occupies the walls of the ventral part of the third ventricle. The **optic tracts** are displaced laterally toward their ultimate termination in the lateral geniculate nucleus in the posterior thalamus. In the temporal lobe the **amygdala** lies deep to the rostral portion of the **parahippcampal gyrus.**

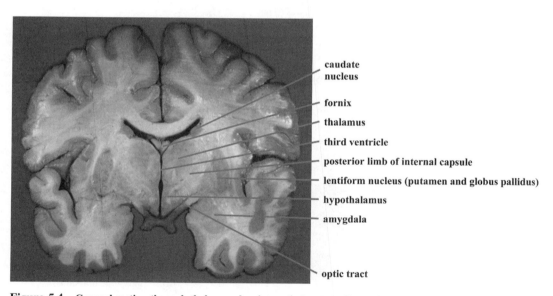

caudate
nucleus

fornix

thalamus

third ventricle

posterior limb of internal capsule

lentiform nucleus (putamen and globus pallidus)

hypothalamus

amygdala

optic tract

Figure 5.4   Coronal section through thalamus showing optic tracts and amygdala.

CORONAL SECTION: THALAMUS AND SUBTHALAMUS

A coronal section more caudally shows the **third ventricle** separating the divisions of the diencephalon. The **subthalamus** lies ventral to the thalamus and lateral to the hypothalamus **(mammillary bodies)** (Fig. 5.5). The **posterior limb of the internal capsule** separates the thalamus from the lentiform nucleus. In the temporal lobe the **hippocampal formation** (hippocampus + dentate gyrus) lies deep to the parahippocampal gyrus, ventromedial to the **temporal horn** (inferior horn) **of the lateral ventricle.**

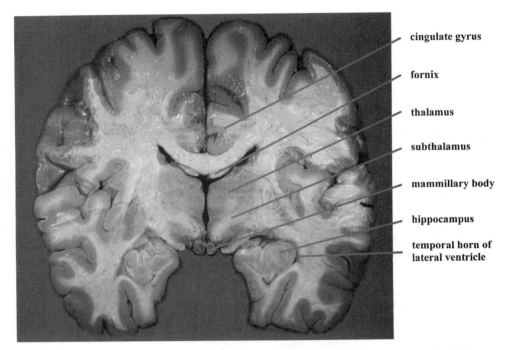

Figure 5.5   Coronal section through middiencephalon showing thalamus, hypothalamus, and subthalamus.

CORONAL SECTION: PULVINAR AND ROSTRAL MIDBRAIN

A coronal section through the caudal thalamus shows the **pulvinar** and **lateral** and **medial geniculate bodies** (Fig. 5.6). The **pineal gland** (part of the epithalamus) lies above the **posterior commissure.** The **hippocampus** is present in the temporal lobe ventromedial

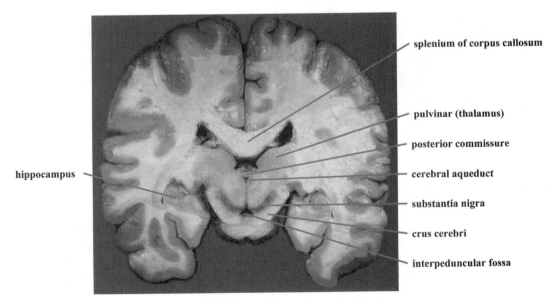

Figure 5.6   Coronal section through caudal diencephalon and rostral midbrain.

to the **temporal horn of the lateral ventricle**. The section traverses the rostral midbrain where the third ventricle is contiguous with the **cerebral aqueduct**. The midbrain tegmentum contains the **red nucleus** and **substantia nigra**. The posterior limb of the internal capsule continued into the **crus cerebri** of the midbrain.

CORONAL SECTION: HIPPOCAMPUS AND ORIGIN OF FORNIX

In a coronal section caudal to the thalamus, the **fornix** can be seen originating from the **hippocampal formation** (Fig. 5.7). The fimbria of the hippocampus becomes the crus of the fornix, and the body of the fornix courses rostrally over the thalamus. The two crura of the fornix are interconnected by the **hippocampal commissure.** On the caudal end of the thalamus the **pulvinar** protrudes above two elevations, the **medial** and **lateral geniculate bodies,** which are relay nuclei in the auditory and visual systems, respectively. Typically, a section at this level also passes through the rostralmost midbrain containing the **superior colliculus, cerebral aqueduct, substantia nigra,** and **crus cerebri** (cerebral peduncle).

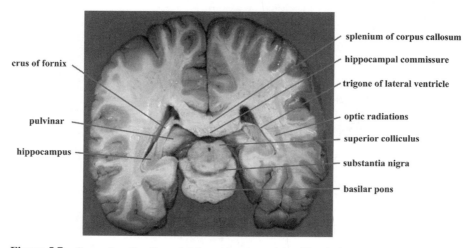

Figure 5.7   Coronal section through trigone showing origin of fornix from hippocampus and midbrain.

ROSTRAL MIDBRAIN

In a **rostral midbrain** section, the **tectum** lies above the level of the cerebral aqueduct and contains the **superior colliculus** (Fig. 5.8). The **cerebral aqueduct** connects the third ventricle of the diencephalon to the fourth ventricle. The **midbrain tegmentum** contains the **oculomotor nucleus, red nucleus,** and **substantia nigra.** Major descending tracts traverse the **crus cerebri**.

CAUDAL MIDBRAIN

In a **caudal midbrain** section, the **inferior colliculus** is in the tectum above the cerebral aqueduct. The core of the midbrain tegmentum at this level contains the **decussation of the superior cerebellar peduncle** where cerebellar efferents cross (Fig. 5.9). Typically, a portion of the **basilar pons** is present in the ventral portion of the bottom of the section.

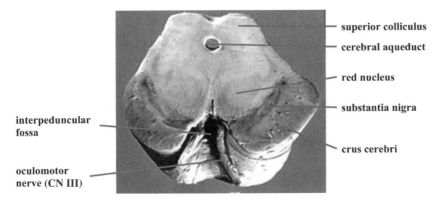

superior colliculus

cerebral aqueduct

red nucleus

substantia nigra

crus cerebri

interpeduncular
fossa

oculomotor
nerve (CN III)

Figure 5.8   Rostral midbrain cross section.

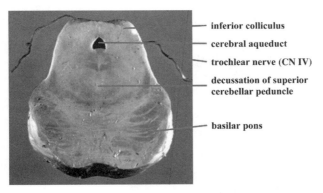

inferior colliculus

cerebral aqueduct

trochlear nerve (CN IV)

decussation of superior
cerebellar peduncle

basilar pons

Figure 5.9   Caudal midbrain cross section with portion of basilar pons.

ROSTRAL PONS

In the **rostral pons**, the two **superior cerebellar peduncles** are connected by the **anterior medullary velum**, which forms a roof over the rostralmost **fourth ventricle** (Fig. 5.10). The pigmented **locus ceruleus** (principal source of norepinephrine in the brain) is seen in the periventricular gray matter. The **pontine tegmentum** is delineated from the **basilar pons** by the flat horizontal sensory tracts, the medial lemnisci. The striated crossing fibers are pontocerebellar fibers that originate from clusters of neurons in the basilar pons (basilar pontine nuclei) and enter the cerebellum through the middle cerebellar peduncle.

CAUDAL PONS

In a section through the **caudal pons** the **fourth ventricle** is near its widest point. Its floor on the dorsal aspect of the **pontine tegmentum** shows the **facial colliculus**, an elevation (medial to the sulcus limitans) produced by the facial nerve coursing over the abducens nucleus. The large **middle cerebellar peduncles** connect the basilar pons to the posterior lobe of the cerebellum (Fig. 5.11). Typically the **trigeminal nerve** (CN V)

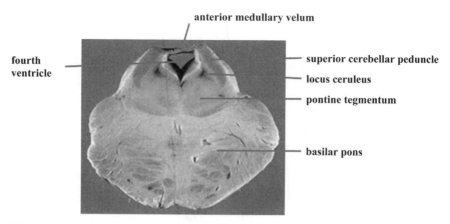

Figure 5.10   Rostral pons cross section.

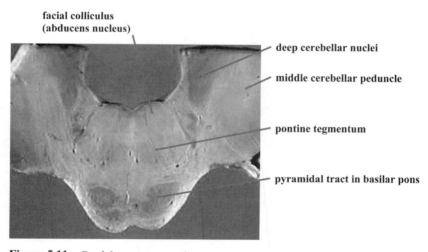

Figure 5.11   Caudal pons cross section.

comes off the peduncle. The large **pyramidal tracts** traverse the **basilar pons** en route to the pyramids of the medulla.

ROSTRAL MEDULLA AND DEEP CEREBELLAR NUCLEI

In the **rostral medulla** the fourth ventricle is at its widest and extends through the **lateral recesses** through the **foramen of Luschka** into the cisterna magna (the largest cistern of the subarachnoid space). These openings along with the midline **foramen of Magendie** are the only passages for CSF to move out of the ventricular system into the subarachnoid space (Figs. 5.12a,b). The **vestibulocochlear nerve (CN VIII)** courses over the dorsal aspect of the **inferior cerebellar peduncle** and ends in an elevation over the cochlear nucleus. The inferior cerebellar peduncle carries tracts from the medulla into the cerebellum. The **inferior olivary nucleus** lies deep to its surface landmark, the **olive**. The **glossopharyngeal (CN IX) and vagus (CN X) nerves** exit from the postolivary sulcus. The

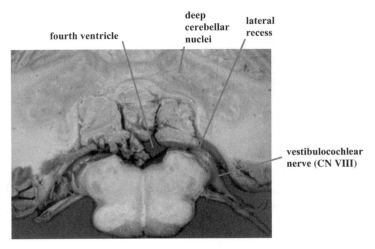

Figure 5.12*a*   **Rostral medulla cross section at level of lateral recesses, showing deep cerebellar nuclei.**

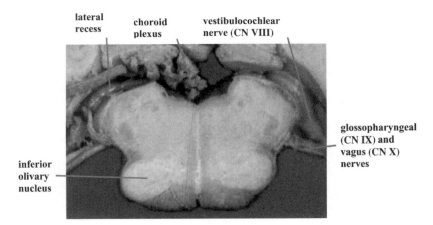

Figure 5.12*b*   **Rostral medulla cross section.**

pyramidal (voluntary motor) **tracts** run deep to their surface landmarks, the **medullary pyramids**.

The cerebellum occupies the roof over the fourth ventricle. It has a midline portion, the **vermis**, and lateral **hemispheres.** Buried within the subcortical white matter, the **deep cerebellar nuclei** (fastigial, globose, emboliform, dentate) are the origin of cerebellar efferents that form the superior cerebellar peduncle.

CAUDAL MEDULLA

In the **caudal medulla** the fourth ventricle has narrowed down to the **obex** (Fig. 5.13). A triad of sensory nuclei occupy the dorsal and dorsolateral medulla: the **nucleus gracilis** (under the **gracile tubercle**), the **nucleus cuneatus** (under the **cuneate tubercle**), and the **spinal tract** and **nucleus of the trigeminal nerve** (under the **tuberculum cinereum**).

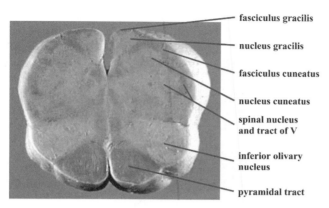

Figure 5.13   Caudal medulla cross section.

The inferior olivary nucleus and pyramidal tracts are seen in the ventral portion of the medulla within the medullary pyramid.

HORIZONTAL SECTION: LIMBS OF INTERNAL CAPSULE

Among the most important horizontal sections of the brain is one where the plane passes through the **internal capsule** showing its three subdivisions: the **anterior limb, genu,** and **posterior limb** (Fig. 5.14a,b). The anterior limb separates the **head of the caudate nucleus** from the **putamen.** The posterior limb separates the **thalamus** from the lentiform nucleus (**putamen + globus pallidus**). The **genu of the corpus callosum** innerconnects

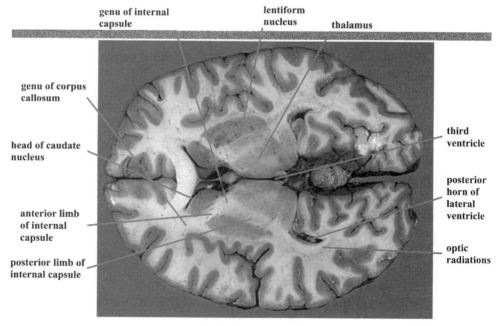

Figure 5.14a   Horizontal section through internal capsule.

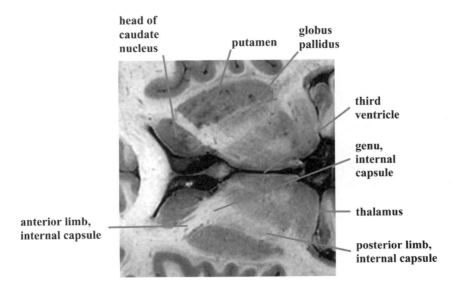

Figure 5.14*b* Basal ganglia and internal capsule.

the frontal lobes, and the **splenium of the corpus callosum** interconnects the occipital lobes. The **lateral ventricles** on either side of the septum pellucidum connect to the **third ventricle** (between the thalami) through the **interventricular foramen of Monro**. The **optic radiations** carrying fibers from the lateral geniculate nucleus to the **cuneus** and **lingual gyri** on the medial aspect of the occipital lobe course in the lateral wall of the **posterior horn** (occipital horn) **of the lateral ventricle**.

Chapter 6

Introduction to Brain Imaging/MRIs

Magnetic resonance images (MRIs) are of such a quality that they superbly picture the sectional anatomy of the brain studied in the previous chapter. The manipulation of the physical properties of the apparatus allows the clinician to better visualize and delineate the borders of lesions or tumors. The T1 MRI has the customary gray/white appearance of the brain, whereas the T2 MRI provides a negative image. In this program MRIs in the coronal plane are emphasized, along with a small number of horizontal MRIs to illustrate the limbs of the internal capsule and cross sections of the brainstem.

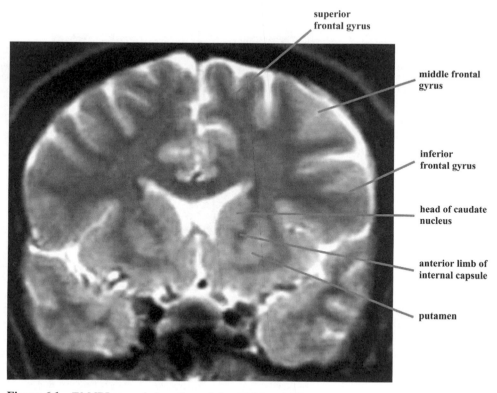

Figure 6.1 T2 MRI coronal plane through frontal lobe, striatum.

T2 CORONAL MRI: STRIATUM

In a coronal MRI through the frontal lobe, the **superior, middle,** and **inferior frontal gyri** are seen. The **cingulate gyrus** lies above the **corpus** callosum. The **head of the caudate nucleus** is separated from the **putamen** by the **anterior limb of the internal capsule**. The **anterior horns of the lateral ventricles** are separated by the **septum pellucidum** (Fig. 6.1).

T2 CORONAL MRI: ANTERIOR COMMISSURE

In a coronal MRI at the level of the **anterior commissure**, the **caudate nucleus** in the lateral wall of the lateral ventricle is separated by the **internal capsule** from the **putamen** and **globus pallidus** (lentiform nucleus). The **optic chiasm** lies below a small portion of the **third ventricle** (Fig. 6.2).

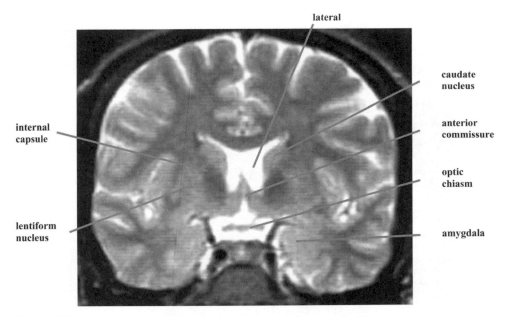

Figure 6.2   **T2 MRI coronal plane through anterior commissure, optic chiasm.**

T2 CORONAL MRI: LENTIFORM NUCLEUS AND AMYGDALA

In a coronal MRI through the rostralmost diencephalon a small portion of the **thalamus** and **hypothalamus** occupy the walls of the **third ventricle** (Fig. 6.3). The **body of the caudate nucleus** is separated from the **putamen** and **globus pallidus** (lentiform nucleus) by the **posterior limb of the internal capsule**. In the temporal lobe, the **amygdala** is medial to the **temporal horn of the lateral ventricle**.

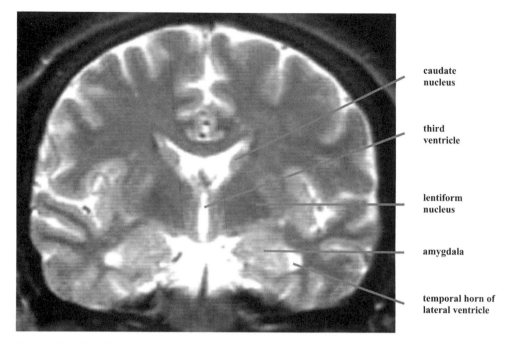

Figure 6.3   T2 MRI coronal plane through lentiform nucleus, amygdala.

T2 CORONAL MRI: THALAMUS, MIDBRAIN, AND PONS

In a coronal MRI through the middiencephalon, the two **thalami** occupy the walls of the **third ventricle** (Fig. 6.4). The lateral ventricles are connected to the third ventricle through the **intervertebral foramina of Monro**. The **posterior limb of the internal capsule** is contiguous with the **crus cerebri of the midbrain** into the **basilar pons**.

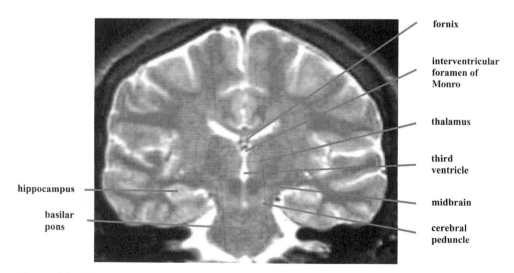

Figure 6.4   T2 MRI coronal plane through thalamus, midbrain, and pons.